Liquid Crystals

Liquid Crystals

Benjamin Outram

IOP Publishing, Bristol, UK

ISBN 978-0-7503-1362-9 (ebook)
ISBN 978-0-7503-1363-6 (print)
ISBN 978-0-7503-1364-3 (mobi)

DOI 10.1088/978-0-7503-1362-9

Version: 20180901

IOP Expanding Physics
ISSN 2053-2563 (online)
ISSN 2054-7315 (print)

British Library Cataloguing-in-Publication Data: A catalogue record for this book is available from the British Library.

Published by IOP Publishing, wholly owned by The Institute of Physics, London

IOP Publishing, Temple Circus, Temple Way, Bristol, BS1 6HG, UK

US Office: IOP Publishing, Inc., 190 North Independence Mall West, Suite 601, Philadelphia, PA 19106, USA

In memory of Oliver Coles.

Contents

Preface

There is profound beauty in observing and understanding the nature of which we are a part, and this book is as much about bringing this into focus as it is about presenting science. The book is aimed at the general reader with an interest in science, and does not presume a science-based higher education. Nonetheless, it does not shy away from discussing the necessary physics, but stops short of presenting any equations. The most casual of readers may only flip through the images, but in doing so will miss out on understanding, which is where the beauty of the images really thrives.

I got involved with liquid crystals quite by accident when my first choice of masters project was not given to me. Instead, my lab partner Graham Renwick and I were given an unlikely collaborative project between the liquid crystal and astronomy departments to create microwave telescopes. Completing that I found myself serendipitously responding to a call by the University of Oxford for a PhD researcher in liquid crystal science.

I was on a landline in a beach hut in Goa, India, when I got the call offering me the PhD position. It is impossible to go to India as a student and not have your mind blown in myriad ways, from the spiritual to the probably-not-spiritual. I have since found the spectacle of physics, nature and liquid crystals to be equally profound. This book is an attempt to share some of that with you.

Acknowledgments

Thanks to the following people for their generous support. My parents, Steve Elston, Stephen Morris, John Sandford O'Neill, Helen Gleeson and her students, John Lydon, Richard Bushby, Flynn Castles, Christopher Robertson, Melissa Lever, Olivia Ferrari, Dan Broughton, Christian Bohley, Andrew Ferris and Max Cooper. Thanks also to the Department of Engineering Science at the University of Oxford and the School of Physics and Astronomy at the University of Leeds who hosted me while I undertook the photography work.

Author biography

Benjamin Outram

Benjamin Outram graduated with a Masters degree in Physics from the University of Manchester in 2009. He completed his doctorate in Engineering Science at the University of Oxford, which included a technology patent, and 12 publications related to liquid crystal science and technology in journals including *Nature Materials, Physical Review E,* and *Applied Physics Letters*. His thesis won the Institute of Physics' Optical Group Doctoral Thesis Prize. Ben received a Daiwa Scholarship and spent four years in Tokyo, Japan, where he completed a Postdoc in Media Technology with a focus on virtual reality at Keio University.

In Tokyo as a Postdoc, he undertook award-winning research in virtual and augmented reality, haptics and computer-human interaction. This resulted in a further 12 publications including one patent related to VR and haptics, and the exhibition of his work at Sundance Film Festival New Frontiers, Tokyo Game Show, Siggraph, IEEE Virtual Reality, ACM Computer-Human Interaction (CHI) and others. Ben is a keen collaborator, having initiated and managed collaborative research between the Universities of Oxford, Cambridge, and Taichung in Taiwan, and Japanese universities and industry, and has secured funding from Merck Ltd, the EPSRC, Daiwa (DAJF) and JSPS. See more from him at his homepage: http://www.BenjaminOutram.com

IOP Publishing

Liquid Crystals

Benjamin Outram

Chapter 1

Introduction

What is it about nature that we find so beautiful? Maybe it is its colours and compositions, or the emotions and thoughts it provokes. This book will explore not only the vibrant visual aesthetic of liquid crystals, but also the deeper beauty that emerges as we come to understand their science and involvement in life.

The often untold story of liquid crystals lies somewhere between chemistry and biology. Most people are familiar with phase transitions like those between ice and water, and water and steam. In some materials, which are common in biological systems, there exist extra phases of matter called liquid crystals. Unlike crystals or liquids, they have a unique combination of fluidity and structure that enables them to spontaneously self-assemble into complex forms while remaining sensitive to environmental changes—key features of biological machinery—with mesmerising visual effect when viewed under a polarising optical microscope. Their fluid clockwork has been critical to the functioning of every biological cell since life began. The lipids, proteins, DNA and chromosomes in your body silently play out their dramas hundreds of billions of times every day.

Considering the shape-shifting, dynamic forms of liquid crystals when viewed under a microscope, it is not surprising that early researchers thought they may be some form of life. Over the last handful of generations, scientists have uncovered many different types of liquid crystals. We have come to understand why they exist, the laws that govern how they interact with themselves, with other materials, with light, and with fields. We have found them in the silk dope of spiders and worms, in the membranes of biological cells and many other biological systems. Chemists have engineered liquid crystals to exhibit new and unusual physical properties. And we have exploited liquid crystals in technology, mainly, thus far, for liquid crystal displays (LCDs) such as those in television and smartphone screens. It turns out that evolution has also worked out how to use liquid crystals for display purposes, as in the iridescence of beetles. We, however, have only just begun to understand the

doi:10.1088/978-0-7503-1362-9ch1

wider implications of imitating how biology exploits liquid crystals for our own technology.

Most people first hear the term 'liquid crystal' in relation to LCDs. Under normal circumstances, the kinds of liquid crystal structures useful for technology tend to be uniform, controlled, stationary, and boring. Some, out of curiosity, may have poked the screens of computers and calculators and been delighted by the strange liquid-like behaviour they saw. By making them flow, a fascinating aspect of their behaviour is revealed. The pretty structures you see throughout this book have come out of the same impetus of curious exploration, taking liquid crystals into conditions that are outside of their use in technology. Messing around with floating pools of liquid crystal near their melting temperature, inducing flow, adding additional chemicals like detergent, and playing with unusual liquid crystal phases—misspent hours of mucking around doing things you're not supposed to just for the thrill of seeing nature doing something intricate and amazing.

In chapter 2 we will build a basic understanding of matter, from the sub-atomic world of quarks to the solids, liquids and gases we experience in everyday life, and we will come to understand what a liquid crystal is. Chapter 3 will introduce light, and how it interacts with matter. These first chapters will form the basis upon which we will understand how matter can produce the images in this book. The remainder of the book chapters will explore the unique features and visual qualities of a series of different liquid crystals. Chapter 4 will explore nematic liquid crystals. The molecular properties that lead to the special physical properties of nematics are discussed. We will introduce key concepts related to the topology and symmetry of liquid crystals, and describe how they are used in LCD technology. Chapters 5 and 6 will introduce the concept of chirality in liquid crystals and explore a range of phenomena related to the twisted structures they form at different length scales. Chapter 7 follows, describing a range of less common and more complex structures, including the cubic-structured blue phases, the lyotropic liquid crystals most commonly seen in biological systems, and the layered smectic and twist-grain-boundary phases, and we will describe how spiders exploit liquid crystals to produce silk with extraordinary physical properties. Finally, chapter 8 will explore the discotic and columnar liquid crystal phases. On our journey, we will learn how liquid crystals interact with light and electric fields, the wonderful structures that liquid crystals can form, and some ways in which biology exploits the unique properties of liquid crystals.

Chapter 2

What is a liquid crystal?

2.1 Protons, electrons and atoms

The stuff of which we are made can be described at a fairly fundamental level as consisting of atoms, which are made of protons, neutrons and electrons. We know that protons and neutrons are themselves made of the same fundamental things, called quarks. These quarks have a property called electric charge, which means that they both produce, and are affected by, an electric field. A field can be thought of like a hilly landscape. Balls roll down to the bottoms of the hills and stay there unless someone gives them a big enough kick to get over the bump to the next valley. The landscape might have high areas and low areas, and the same is true of electric fields.

Now, imagine if the landscape were actually a humongous rubber sheet, and the balls on the sheet were big, heavy steel balls. The balls would make big indentations in the sheet, thereby themselves affecting the landscape. If two such heavy balls existed on this sheet, they would roll inwards towards each other, down in to the valleys caused by their own weight. This analogy is often used to describe gravity. Objects with mass have a gravitational field that attracts other objects with mass. But charge is different. There is, as far as we know, only one type of mass, and one type of gravitational field. But charge comes in two types, positive and negative. Opposite charges are attracted to each other, and like-charges repel.

As well as protons and neutrons that are made of quarks, atoms also have electrons. If you take an atom as a whole, it doesn't have any charge. This is because the charge of all the electrons is exactly equal and opposite to the charge of all the protons in the atom. For example, an atom of hydrogen has one proton, which we say has a charge of +1, and one electron, having a charge of −1, resulting in a total charge of zero.

Protons and neutrons are made of three quarks each. The two types of quarks that exist in protons and neutrons are called 'up' and 'down' quarks, and they have charges that are a fraction of the electron charge, but when they come together as a group of three, their charges add up to exactly zero in the case of a neutron, and

doi:10.1088/978-0-7503-1362-9ch2

exactly +1, in the case of a proton. Why do quark charges have the perfect fraction of charge to perfectly counter the electron charges to make neutral atoms? This is still a profound mystery.

As mysterious as particle charges and atoms are, perhaps more mysterious is the way electrons behave around and between atoms. In school we usually draw the protons and neutrons in the atom's nucleus and then draw rings around in which electrons are orbiting, a bit like planets around the Sun as in our solar system, but the truth is far more interesting. It turns out that mechanics at the scale of atoms is very different from the way we normally think about matter interacting.

Quantum mechanics is the most useful theory for describing small things like electrons in atoms. Using quantum mechanics, physicists have been able to describe something about the nature of electrons in atoms, and it's not like planets orbiting a star at all. If you have a lone electron travelling around in space, it can be described as a wave, a bit like a wave moving on the surface of water. When the electron is trapped by the positive electric field created by the quarks in an atomic nucleus, the wave gets trapped, a bit like the wave trapped in a vibrating guitar string. Because a string is a particular length and tightness, it vibrates at particular frequencies. The electron wave does a similar thing around the nucleus of an atom. They have a stable state in a spherical orbital shape around the nucleus. Other shapes are possible too, and if an electron gets excited, like plucking a guitar string, it can jump between shapes of tiny spheres, loops and rings.

Later we will see that light, which is a kind of electromagnetic field, can affect and be affected by the electrons in atoms and molecules, and it will be useful not to think of atoms as mini-solar systems but as dynamic wavy-fieldy things that interact with and cause fields. Liquid crystals, and life, exploit the complex interactions possible in this wonderful world of forces and particles.

2.2 Molecules

In order to understand liquid crystals, we need to comment on how atoms can organise themselves into molecules, and molecules into phases of matter such as liquids, solids and gases.

Although atoms are neutrally charged overall, they commonly bunch together to form molecules. In a molecule, one or more electrons' orbital shapes are shared across two or more atoms. In such a scenario, we have a neutrally charged molecule consisting of two or more atoms, with some of the electrons orbiting around two atomic nuclei, and some just around one (figure 2.1).

Depending on how electrons are shared and distributed, electric charge within the molecule can become concentrated in some areas more than others. This sets up internal forces that result in molecules with certain preferred shapes and angles between the constituent atoms. For example, the angle formed by the oxygen and the two hydrogens that form water is 104.5°.

The mobility of the electrons within organisations of atoms can also vary. For example, in metals, some of the electrons offer little resistance and can flow through the material. Some molecular structures also allow electrons to move more easily

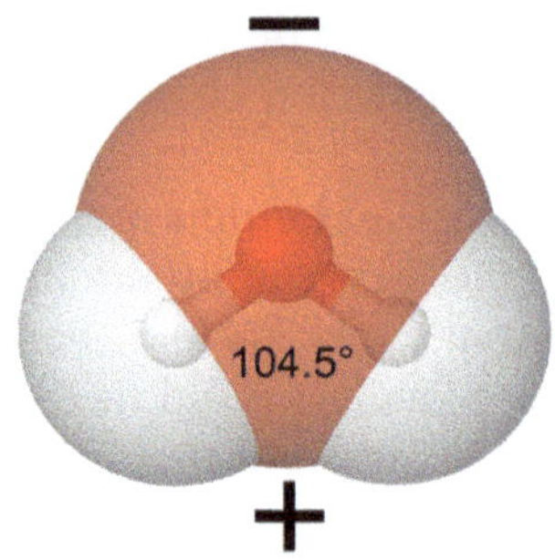

Figure 2.1. A representation of a water molecule, H_2O. The two hydrogens (white) share electrons with the oxygen (red). The negative charge of the electrons on the far side of the oxygen push the hydrogens to one side, resulting in positive and negative ends.

around the molecule, for example, a ring of carbon molecules, as is very common in biological materials, has a relatively large electron mobility within and between adjacent rings. We will see later how this influences how light interacts with liquid crystals and is exploited for liquid crystal displays.

The way that all this charge is distributed and can move around also determines how the molecules interact with each other, resulting in the variety of phases of matter that we know of, including solids, liquids, gases... and liquid crystals. In addition, it leads to 'amphiphilic' and 'hydrophilic' behaviour, or attraction to fat or to water, which are critical to the further organisation of phases of matter into structures such as cell membranes, and beyond.

2.3 Phases of matter

From the mysteries of quarks and electron orbitals, we have come to the organisation of matter into molecules. Now we will consider what happens when molecules interact with each other to form the phases of matter we are familiar with in our everyday life, such as solids, liquids and gases.

Since the total charge of a molecule is zero, you may be led to think that there are no electric forces between them. Indeed, this is a good assumption if they are sufficiently separated. However, since the charge distribution within molecules is not the same all over the molecule, attractive and repulsive forces can and do come into play when the molecules are close together. It is the net effect of these intermolecular forces that determine the phase behaviour of a material—solid, liquid, gas—and liquid crystalline.

As we know from our experience with freezing and boiling water, the same material can exhibit many different states of matter, depending on the temperature and pressure. Molecules of water consist of an oxygen and two hydrogen atoms, but the hydrogen atom's electrons are pulled over towards the oxygen molecule. Since the hydrogens form an angle of 104.5° with the oxygen, the molecule's charge distribution is weighted such that the charge is negative on the side of the oxygen, and positive on the side of the hydrogens. This is called a dipole—which means 'two poles'—because the molecule has a positive pole and a negative pole. When two water molecules come close together, the oppositely charged sides of the molecules flip around and are attracted towards each other. If multiple water molecules are present, these attractive forces result in the formation of a crystal of water ice.

The molecules in ice can form a variety of different highly ordered crystal structures, depending on their initial conditions and how they froze. We know of more than 16 different crystal structures of water ice and new ones are occasionally still discovered! It seems we still have a lot to learn about even water, although almost all of the water ice on earth is of just one type, forming a hexagonal lattice structure.

As ice warms up, the heat energy in the crystal lattice increases. You can imagine the atoms fixed in their molecular formations, and the molecules fixed in their crystal lattice formations, but nonetheless jiggling about. The atoms don't stay perfectly still, because there is always some kinetic energy with nowhere to go. If this energy is small, the molecules stay in their crystal lattice positions. But as we add heat to the system, at some point the molecules have enough energy to break free of the attractive forces between themselves. At this point, additional heat no longer increases the vibrational energy of the molecules in their lattice positions, but instead is used to break up the lattice structure resulting in the phase transition from crystalline ice to liquid water, which we call melting.

In liquid water, there is too much thermal energy for molecules to settle to form fixed lattice structures, but molecules still experience the attractive forces between their oppositely charged sides. Whereas in a crystal, molecules have fixed positions relative to the lattice of molecules in the crystal structure, in a liquid, molecules are able to move around. This is why water can flow easily, and ice cannot.

While we know that ice snowflakes have six sides and therefore a six-fold rotational symmetry—which is due to the underlying hexagonal crystal lattice structure—liquid water appears the same no matter what angle, or in what position, you examine it. In other words, undergoing a phase transition often has the effect of changing the level of symmetry of the material, in this case rotational symmetry and translational symmetry. Rotational symmetry means that the structure looks the same when viewed from different angles. Translational symmetry means it looks the same from different positions. This will be particularly important when we consider related transitions between phases of liquid crystal.

As we know, water undergoes another phase transition at 100 °C, from liquid into gas, which we call boiling. If you measure the temperature of water as you bring it to the boil, you will find the temperature rising and rising until it reaches 100 °C, at which point the temperature will not rise any further. The random jostling of molecules becomes so severe that the intermolecular attraction is not sufficient to keep molecules from separating from each other. The heat energy that you continue to put into the water from the burning gas in your hob, rather than increasing the temperature, goes into breaking those attractive forces and turning the water into steam.

The thermal energy in steam prevents the molecules from sticking together. Molecules will happily fly out and away until they hit something, such as the wall of the container that they are in, like balls on a pool table after a strong break. The only reason the atmosphere does not escape into deep space, is because earth's gravity is sufficient to make molecules fall back to earth should they find themselves hurtling upwards.

What happens when we add even more heat to our steam? There are no more intermolecular bonds left to break, and the temperature of the steam rises. The thermal energy goes into increasing the speed of the molecules, and they fly around hitting each other and the walls of the container with ever greater speed. Some of this energy also goes into vibration of the atoms about their fixed positions in the molecules. Indeed we still have electrons that are attached to their atoms, and atoms connected via intramolecular bonds to each other, but at some temperature even these bonds can break. The energy to liberate an electron from an atom is called the ionisation energy. If the thermal energy surpasses this energy, the material may undergo a further phase transition to a plasma—a gas of ions, or charged particles.

If the thermal energy surpasses the bond energies holding the molecule's atoms together, it is possible for these to break as well. This is called thermal decomposition. Unlike a true phase transition, a thermal decomposition is not usually reversible—the atoms will not in general recombine if you cool the material down again. Which happens first, ionisation or decomposition, or indeed decomposition or boiling, depends on the relative strengths of the forces between atoms in molecules, forces between molecules in liquids and solids, and the ionisation energies. A textbook example of thermal decomposition is from calcium carbonate (chalk or limestone) into calcium oxide (quick lime) and carbon dioxide, which happens above 825 °C. Water does not decompose until it reaches temperatures of well over 2000 °C.

So, you may be asking, in this picture of phases of matter, where do liquid crystal phases come?

2.4 Liquid crystal

Imagine now that instead of water ice, we have a crystal of a material made out of the molecule shown in figure 2.2. Notice that, unlike water molecules, this one has a long shape.

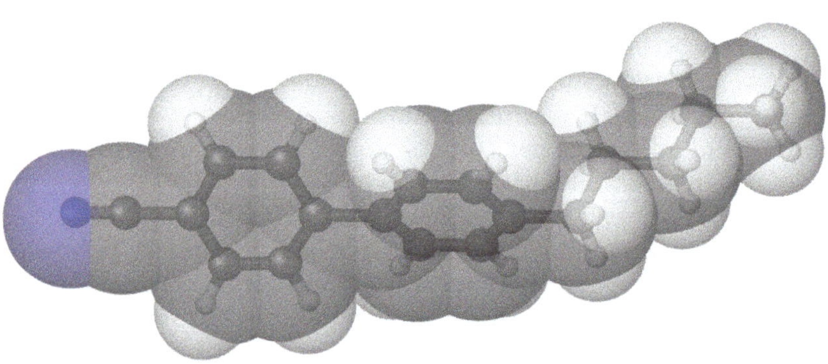

Figure 2.2. A diagram of a 5-cyano-biphenyl (5CB) molecule, which is a common component of liquid crystal materials. Blue represents a nitrogen atom, and grey and white are carbon and hydrogen, respectively. The elongated shape is critical to its liquid crystal phase behaviour.

In order to understand the behaviour of such a molecule, imagine a bucket filled with matches, which represent these long thin molecules. When still, the matches do not move relative to each other, a bit like molecules in a crystal. As we start to jostle the bucket, the matchsticks also begin to jostle. This is like adding heat energy to our material. At a certain level of jostling—imagine jostling it quite hard and fast—the matchsticks will no longer stay in a fixed position, but begin to move about, never staying in one place. This is like a phase transition of our material from crystalline into a fluid. But look at your imaginary bucket of matchsticks and what do you see? All the matchsticks, jostling about, but they are all pointing roughly in the same direction! Some areas point one way, and other areas point in others; matches at the edge tend to lay flat against the side of the bucket, not always, but in most places. But for sure, matches tend to face in a similar direction to ones directly next to them. What you are imagining is a nematic liquid crystal phase. The direction in which the matches tend to align is called the 'director'.

Water is isotropic, which means that it has the same physical properties and looks the same no matter which angle we look at it from. Liquid crystals such 5CB on the other hand are anisotropic—they have a direction and therefore different properties along different directions.

It is their fluidity and anisotropy that give liquid crystals their useful physical properties. As we will see over the next chapters liquid crystals have anisotropic optical, electric, viscous and elastic properties, all of which are exploited by biological and technological liquid crystal machines.

Let us make one more visit to our imaginary bucket of matchsticks. You were last imagining shaking the bucket around to simulate thermal energy, which allowed the matches to move freely while staying oriented along a common director. Now let's go further. Put the lid on the bucket lest they escape and start violently shaking the bucket even harder and faster. This simulates heating up the liquid crystal to a higher temperature. Enough energy, and even the matches' orientations will be

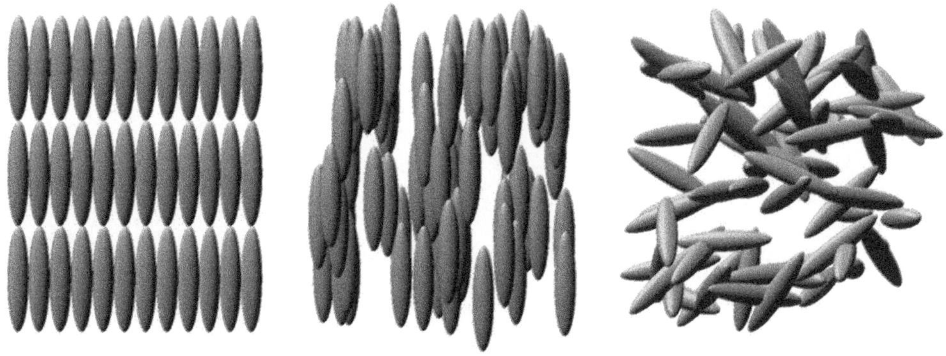

Figure 2.3. Three phases of matter exhibited by elongated molecules like 5CB. At low temperatures the material forms a crystal structure, shown in the left image by a regular array of elongated shapes that represent molecules. At warmer temperatures, the crystal undergoes a transition to the nematic liquid crystal phase in the center, where molecules are on average pointed in a particular direction. If there is enough heat energy, even this nematic ordering gives way to an isotropic symmetry, with molecules randomly oriented as in the right hand image.

jostling randomly around, no longer exhibiting a preferred direction. At this point, the material loses its anisotropy, and undergoes a phase transition from a liquid crystal to an isotropic liquid.

Figure 2.3 illustrates three phases of 5CB, going from the most ordered crystal phase, up in thermal energy to the liquid crystal phase, and finally to the isotropic liquid phase. We have spoken here of only one kind of liquid crystal phase, but as we will see in later chapters, there are many others as well.

In this chapter we have seen how quarks and electrons bind together to form (mysteriously) neutral atoms, how atoms bind together to form molecules, and how molecules interact to form phases of matter including solids, liquids, and gases. Finally, we imagined how long and thin molecules could produce anisotropic liquids, which we call liquid crystals. In the next chapters we will begin to understand how liquid crystals interact with light, and how they result in the photographs throughout this book.

IOP Publishing

Liquid Crystals

Benjamin Outram

Chapter 3

Light

3.1 What is light?

We already came across the idea of charge and electric fields. We know that charges are affected by and also create electric fields, and we know electrons have charge. We also came across the idea of an electron as being a wave, either travelling through space or being trapped in an atomic electron orbital by the opposite charge of an atom's nucleus. We know that light can affect and be affected by matter—but light itself does not have any charge. So what is light?

Between the work of the two famous physicists, Maxwell and Einstein, we have discovered that magnetic fields exist because electric charge comes in two types, positive and negative, and that electric and magnetic fields are actually the same thing—just seen from different frames of reference. A consequence of this is the existence of light, which is an electromagnetic wave. The electric field oscillates from positive to negative as the light travels through time and space, mediated by a magnetic field oscillation.

For now, it is enough to understand that light is a wave in space, and because it is an electromagnetic wave, it can and does interact with electrons. While light does not have charge and is therefore not trapped by atomic nuclei like electrons are, its interaction with electrons is responsible for how light is slowed down or absorbed by matter.

Electromagnetic waves come in a wide range of frequencies, and the higher the frequency, the shorter the wavelength. Radiowaves, microwaves, x-rays and gamma rays are all electromagnetic waves of different frequencies. Light that is visible to the human eye is in the range of optical frequencies, with red light having the longest wavelength of 700 nanometers, and violet having the shortest of 350 nanometers, which is about 1/100th of the thickness of a human hair.

3.2 Polarised light

When you drop a pebble into a pond, ripples spread calmly outwards in concentric circles. The waves themselves are characterised by the height difference (amplitude)

doi:10.1088/978-0-7503-1362-9ch3

and separation (wavelength) between the crests and troughs of the moving ripples, and they move in straight lines radially out from where the pebble went plop. They are called transverse waves, because the movement of the water is perpendicular to the wave's movement direction.

Waves of light, on the other hand, are not confined to the surface of a pond, but instead can radiate in any direction in three dimensions. Light is also a transverse wave, but is characterised not by the amplitude and wavelength of the height of a water surface, but by the amplitude and wavelength of the electric and magnetic fields.

Since ripples of water happen on two-dimensional water surfaces, it is clear which direction is up, and therefore the direction in which the water moves up and down. For electromagnetic waves travelling in three-dimensional space, however, there is no such obvious preferred axis along which the electric field changes. This fact leads to a very important feature of light waves, for which there is no analogy in water waves: polarisation.

A light wave coming from this paper and hitting your retina could have a polarisation such that its electric field is oscillating up and down, but alternatively, it might equally well be oscillating from left to right, or indeed at any angle, so long as the direction is at right-angles to the direction the wave is propagating. Figure 3.1 illustrates how the electric field oscillates in horizontal and vertical linearly polarised light.

The polarisation state can affect how light interacts with matter, for example, how it reflects off surfaces such as roads and water. Polarising sunglasses are made of a material that is able to selectively block left–right polarised light and thereby selectively cut out distracting reflections of light from the road. Some animals are capable of detecting light polarisation too, including honey bees and octopuses. They use it to detect water surfaces, and by looking at the light scattered by the sky, as a compass for navigation.

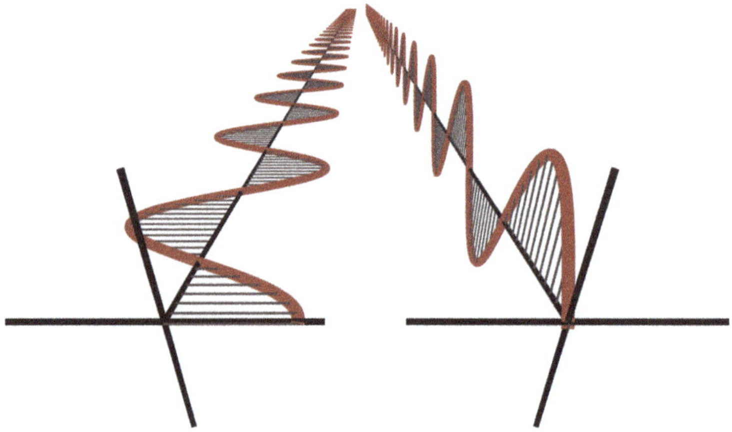

Figure 3.1. The left and right diagrams show the electric field vector of horizontal and vertical linearly polarised light travelling into the distance. The electric field oscillates between a maximum and a minimum. Different coloured light has different wavelengths.

Figure 3.2. The left and right diagrams show the electric field vector of left- and right-handed circular polarised light travelling into the distance. The electric field vector rotates in a spiral as it moves through space.

3.3 Circular polarisation

Imagine that you are holding a rubber band, on the end of which you have tied a small stone. As you shake your hand up and down, the stone springs up and down, its velocity mediated by the oscillating elastic force in the rubber. Equally, you could shake from left to right.

Now imagine that, instead of wiggling the stone up and down, you spin it around in a circle. Notice that, whereas before the speed of the stone oscillated between a minimum and a maximum, now the speed is at a maximum all the time, on a circular path.

Electromagnetic waves have a similar behaviour. The electric field can oscillate up and down or left to right, but can also rotate around in a circle with its magnitude at a maximum all the time. This is called circular polarisation. Figure 3.2 illustrates the electric field in left and right circularly polarised light.

Just as you can spin a stone on elastic around in clockwise and anti-clockwise directions, so too do left and right-hand circular polarisation states of light exist. In addition, you can imagine intermediary states in which the pebble moves in an ellipse. So too do elliptical polarisation states of light exist.

Now that we have some kind of picture of electromagnetic waves and polarisation, we can begin to understand why materials like liquid crystals interact with light in the way they do, providing such pretty colours.

3.4 How is light affected by matter?

In transparent materials such as water and diamond, the existence of charged electrons means that an electromagnetic wave can have the same frequency while having a shorter wavelength. The oscillating electric field interacts with the electron charge, which in turn affects the electromagnetic wave. A consequence is that the wavefront of the light moves at a slower speed than if it were in space. The 'refractive index' of a material is defined by how much the light is slowed down, or 'retarded', which in turn depends on the density and mobility of electrons in the material for a given frequency of light.

If we imagine a light's wavefront coming towards a flat surface of water at an angle, some of the wavefront will enter the material earlier than others, and so be slowed down first. The result is that the wavefront progresses through the material at a different angle than before it entered the material, and the light is refracted, or bent. Shorter wavelengths of light are refracted more than longer wavelengths, which accounts for how the Sun's white light is split into colours at slightly different angles in the sky when it undergoes multiple refractions in droplets of water, resulting in beautiful rainbows. It also accounts for the sparkling colours we see in diamonds.

If the light's photons have energies that are just about right to excite electrons into higher energy states, then electrons tend to absorb the photons and the electromagnetic energy is converted into heat. This means that the intensity of the light decreases as it moves through the material, rather than being transmitted. Diamond and coal are both made of carbon atoms but the electron excitation energies in the materials are different, resulting in diamond being transparent to visible light, and coal absorbing all of it.

3.5 Light and anisotropic materials

Anisotropic materials such as nematic liquid crystals have different physical properties in different directions. Imagine a sheet of the liquid crystal material we described in section 2.4, whose director has been aligned uniformly in a single direction along the sheet. As we have discussed, the refractive index depends on the mobility of the electrons within a material. For anisotropic materials, this mobility is generally different parallel and perpendicular to the molecules.

In our example liquid crystal molecule, 5CB, we have two hexagonal carbon rings. These ring structures contain electrons that are relatively mobile at optical frequencies, but the electrons are more confined in the direction perpendicular to the connected ring structures than parallel to them. The result is that the part of the electromagnetic wave whose electric field oscillates parallel to the nematic director gets slowed down more than the part that is perpendicular. This effect is called birefringence, which means that the material has two refractive indices. The direction parallel to the director is called the 'optic axis'.

Let's return to our analogy of the stone on the elastic band. Imagine again that the stone is oscillating up and down on the springy rubber band. Now imagine that we increase the stiffness of the rubber band. This has the effect of increasing the frequency of the oscillation. We can imagine that to perceive the oscillation at the same speed as before, we would have to slow down time. This is analogous to the retardation experienced by light as it passes into a material. It is as if time is slowed down, with the wave of light moving at a slower speed.

Now let's consider a birefringence in our analogy. Imagine that your stone is on a hypothetical rubber band that has a different elasticity in vertical and horizontal directions. In reality, this is impossible for a rubber band, but let us anyway imagine what the consequence would be. If you shake up and down, the stone will oscillate at one frequency. Shake horizontally, and the stone shakes at a different frequency, which depends on the stiffness of the elastic. We can say that parts, or 'components', of the oscillation that happen in vertical and horizontal directions each experience the corresponding vertical or horizontal 'component' of the rubber band's elasticity.

Now imagine trying to oscillate the stone not in purely horizontal or vertical directions, but diagonally. At first the stone will bounce back and forth on a diagonal axis. It has a component of its movement in the vertical direction and a component in the horizontal direction, but each component oscillates at different frequencies. As time moves on, the horizontal and vertical components become out of phase. As we continue to watch, the stone begins to rotate, first forming an elliptical path, and then a circular one. From a circular path, it will then again become elliptical, before returning to a linear oscillation, this time at 90° from where you began.

This is exactly what happens to the polarisation state of light as it passes through a birefringent material. The components of the wave of electric field parallel and perpendicular to the optic axis of the material travel at different speeds, and so the phase between them changes.

When the light finally leaves the material, the light will continue with whatever polarisation state it was in before leaving the material. It is therefore possible to engineer a sheet of birefringent material that has the perfect thickness to convert a linear polarisation into a circular polarisation, or vice versa! We call this a quarter-wave plate, because the perpendicular and parallel components of the electric field oscillation become shifted in space by exactly a quarter of a wavelength.

Different wavelengths experience different amounts of retardation, which we understood to be responsible for the colours of rainbows and sparkles in diamonds. It is also true that different wavelengths will experience different amounts of shift in their polarisation states as they move through a birefringent material, and it is this effect that is primarily responsible for producing the beautiful colours in the polarising microscope images in this book.

An anisotropic material may also absorb light differently in different directions. Such a material is used to create polarisation filters that only allow one polarisation state through, such as are used in polarising sunglasses. This is the primary way that scientists use to produce polarised light from a source of randomly polarised light, such as from a lightbulb. As we will see in the next section, polarising filters are used in polarising microscopes, which were used to take all of the photos in this book.

3.6 Polarising optical microscopy

A polariser is a filter that only allows one linear polarisation of light through. It works by using a material with an anisotropic absorption, that completely absorbs light polarised along a particular axis as it passes through the material. Now, imagine if we were to have two polarisers and place them both in the path of some light. If we aligned the polarisers so that their axes were aligned parallel, then the light that comes through the first polariser would have the correct polarisation to be able to get through the second one, and the light would be transmitted. On the other hand, if the polarisers were set with their axes perpendicular to each other, then the linear polarisation that comes through the first would be completely absorbed by the second polariser.

In a polarising microscope, two polarisers are placed in the path between the lightbulb at the bottom and the eye-piece or camera at the top. The second polariser that light reaches is called the analyser. Usually, the analyser's axis is set perpendicular to the polariser, so that in the absence of any sample, all the light is absorbed and it appears black down the microscope. Such a setup is called having crossed polarisers.

Normally, crossed polarisers absorb all the light. However, recall that light passing through a birefringent material undergoes a change in its polarisation state. Therefore, the birefringence will enable some of the light to pass through. How much light passes depends on its wavelength, the birefringence of the material, how thick the layer is, and the angle that the optic axis of the material is in relation to the crossed polarisers. We can therefore use polarising microscopes to look at the structure of birefringent materials at the microscopic scale, which by the way, is exactly the scale at which liquid crystalline structures tend to form.

Chapter 4

Nematic liquid crystals

Even the simplest of liquid crystals, called nematics, produce intricate beauty when observed under a polarising microscope. Notice how the structures in images 4.2–4.4 follow specific patterns. The smoothness of the gradients are counterpointed by a variety of sharp pinwheels that betray some underlying order or topology. This order comes from the behaviour of the nematic director (the prevailing local orientation of the liquid crystal molecules), which we can manipulate using heat, flow, and chemicals to craft the images in this chapter.

4.1 The nematic liquid crystal molecule

There are many molecules that form a nematic liquid crystal phase. Let's look at the 5CB molecule again and recap the key concepts we have covered so far, which ultimately lead to the behaviour we see in these images (figure 4.1).

- A hydrocarbon is a molecule or part of a molecule that consists of a chain of carbon atoms covered in hydrogen atoms. The hydrocarbon chain on the 5CB consists of five carbons and produces most of the length of the molecule. It is the elongated shape, or shape anisotropy, that results in the macroscopic anisotropy of the material.
- A phenyl ring is a ring of six carbon atoms. The two phenyl rings form a biphenyl ring complex. These structures would not be stable if it were not for the overlap of the electron orbitals in the centre of the rings. The electrons in these rings have a higher mobility between the rings than across them, which contributes to their important anisotropic refractive index, or birefringence. This is important for allowing them to interact with polarised light so that they can be studied under a microscope and exploited to control light by liquid crystal displays.
- The nitrogen at the end of the molecule shares not one but three electrons with the carbon next to it. This is called a cyano group (although in this case it is not poisonous like cyanide). The cyano group tends to pull negative

doi:10.1088/978-0-7503-1362-9ch4

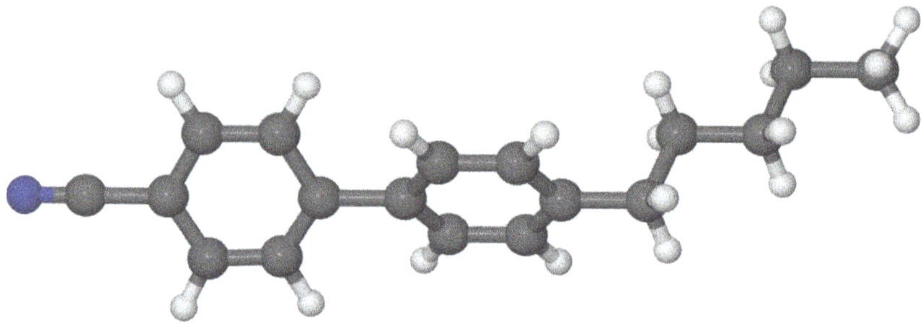

Figure 4.1. The 5CB molecule has three primary functional components. Its shape anisotropy, mostly due to the hydrocarbon chain on the right, results in macroscopic anisotropy of the phase. Its biphenyl ring causes anisotropic dielectric permittivity at optical frequencies. Finally, the electronegative nitrogen enhances the electric dipole moment, enhancing the molecule's interaction with static electric fields.

charge towards that end of the molecule, while the hydrocarbon chain tends to push electrons. The result is an electric dipole along the molecule. The dipole allows the orientation of the director to be controlled by applying an electric field.

4.2 Director orientation gradients

We already came across thermotropic nematic liquid crystals when we imagined the phase behaviour of long and thin molecules of 5CB in section 2.4 using the analogy of matchsticks in a bucket. A nematic liquid crystal is one in which molecules tend to align along a director, and 'thermotropic' means that it can melt into a normal isotropic liquid if heated. Imagine we cool a bottle of this material from its isotropic phase down into its nematic liquid crystal phase. The molecules, being randomly oriented in the isotropic phase, suddenly align with their neighbours in a uniform direction, the local director. The question is, which way will the director end up facing?

The answer is that there is no particular preferred direction. At the moment of becoming a nematic phase, different regions of liquid crystal will have their director pointed in different directions. The long molecules continuously jostle about, bumping and influencing molecules next to them. Our bottle of liquid crystal might be a centimetre in diameter, but liquid crystal molecules are only of the order a nanometer (10^{-9} m) long. The width of our liquid crystal bottle is approximately ten million molecule-lengths! Forget a bucket of matchsticks; one of Canada's Great Lakes filled with matchsticks would be a more appropriate image. Since molecules prefer to lie parallel to molecules next to them, over time regions of uniformly aligned director tend to grow, but the entire bottle will never become aligned completely uniformly.

In the isotropic phase, the liquid is transparent like water. However, the nematic phase is birefringent—light is slowed down differently depending on the orientation of the director. Since different regions in the bottle are oriented in different directions the light is scattered through refraction, giving bottles of nematic liquid crystal a milky appearance.

Under a polarising microscope, a film of isotropic liquid appears black because the crossed polarisers block all of the light. However, birefringence of the nematic liquid crystal phase affects the polarisation of the light, resulting in an image. Figure 4.2 shows a thin layer of nematic liquid crystal in the most iconic of all types of liquid crystal formation. Known as the Schleiren texture, the striking gradients of coloured light and dark are smooth except for occasional sharp singularities all over the image.

The smooth gradients correspond to areas where the director is fairly uniform. Recall from chapter 3 that light will undergo the greatest change in its polarisation if its electric field oscillation is diagonal to the liquid crystal director, allowing light to pass through the microscope's analyser. This corresponds to the bright regions of the image. Then, where the director is oriented parallel or perpendicular to the illuminating light's polarisation corresponds to the black regions.

In chapter 3 we also mentioned that the amount of change in polarisation state of light passing through a birefringent material such as liquid crystal depends on the wavelength of the light. This accounts for the different colours that we see in the images in this chapter. A very thin layer of liquid crystal will appear grey or white. Longer wavelengths are affected more, so slightly thicker layers appear yellow. As the layer gets thicker, other colours are possible; red, magenta, blue and green, depending ultimately on the polarisation state of all the frequencies of light exiting the liquid crystal layer. The colour therefore reveals information about the birefringence and thickness of the liquid crystal.

4.3 Liquid crystal defects

Referring again to images 4.2–4.4, we see places where the continuous, smooth gradients come to a point, always with four light brushes radiating out. These are

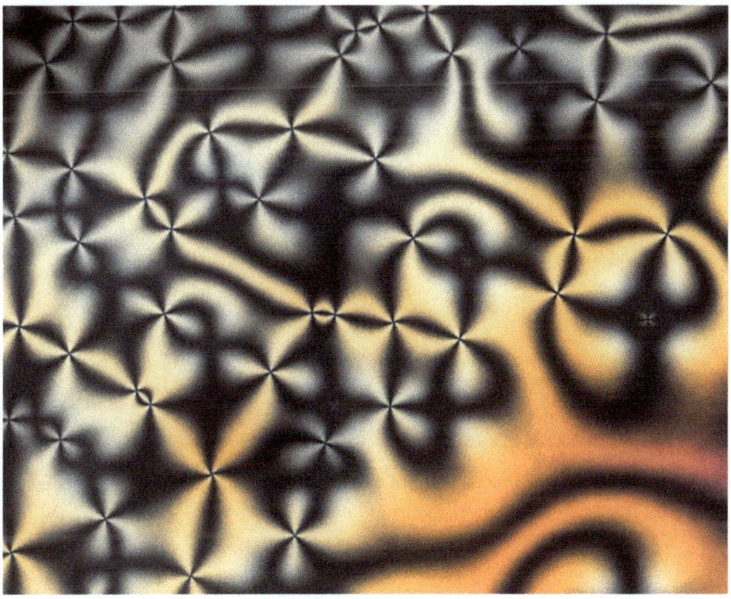

Figure 4.2. A Schlieren texture of the nematic liquid crystal phase.

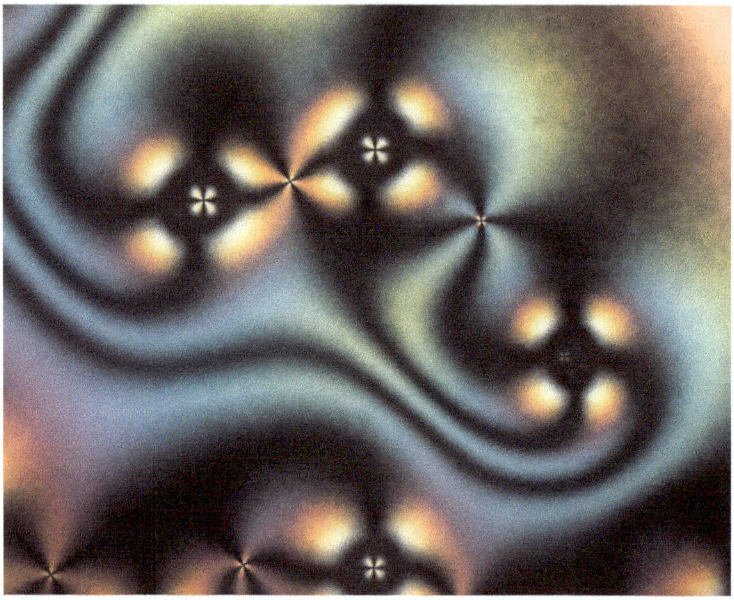

Figure 4.3. Defects seen in a layer of nematic liquid crystal floating on a layer of isotropic bellow.

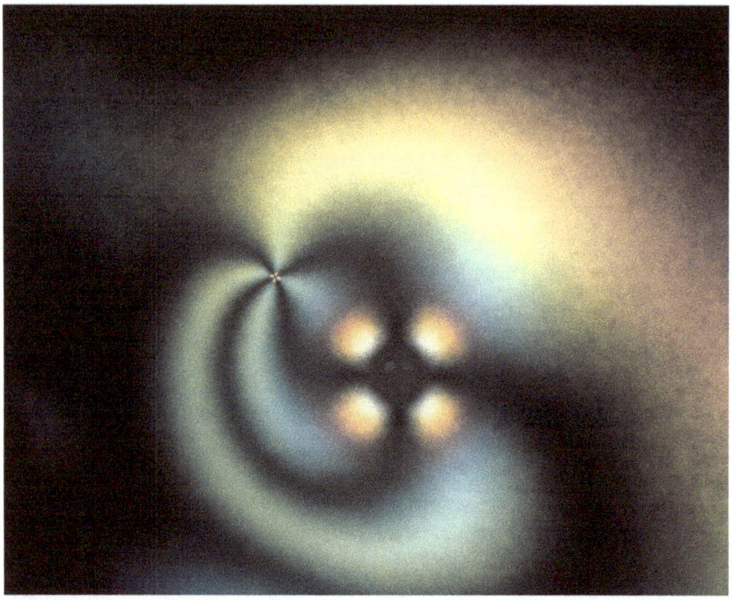

Figure 4.4. A +1 and −1 defect pair in a layer of nematic liquid crystal close to the nematic-to-isotropic transition temperature.

discontinuities in the director that arise due to a conflict between the liquid crystal director's tendency towards uniformity, and the local director topology which prevents it.

Figure 4.6 illustrates a few ways in which the director can be oriented around such discontinuities, which are also known as liquid crystal defects. Defects are classified by how the director orientation changes as you trace a line around the defect. If the director rotates once it is called a +1 defect; if it rotates the opposite direction, it is called a −1 defect; if it only rotates 180 degrees, it is called a +0.5 defect.

Figure 4.5 illustrates the director structure overlaid on a polarising microscope image of a nematic liquid crystal. Notice that the director is at plus or minus 45° with respect to the polariser when the image is brightest, and black where the director is perpendicular or parallel to the polariser. As we go around a defect, the director makes a diagonal with the polarisers four times as it rotates through 360°, which explains why the defects seem to have four dark brushes coming out from them. Look back at figure 4.2 and see if you can imagine how the director of the nematic liquid crystal is oriented around the image. Although we cannot see the director structure directly, we can see the effect of birefringence on the polarised light.

In order to distort the nematic director away from being in a uniform, single direction, it costs energy. This is analogous to the effort you have to put in to bend something springy. In figure 4.6 we can see that the director around defects is highly distorted, and this means they contain a high elastic energy. +1 and −1 defects have opposite types of distortion, and when they come together they merge and annihilate to become a uniform director. If there are many defects present, for example, just after a phase transition from the isotropic phase, then opposite defects will quickly merge together and over time become fewer, reducing the total elastic energy in the liquid crystal system. In some cases though, defects may spontaneously form if doing so reduces the overall energy of the system in other ways.

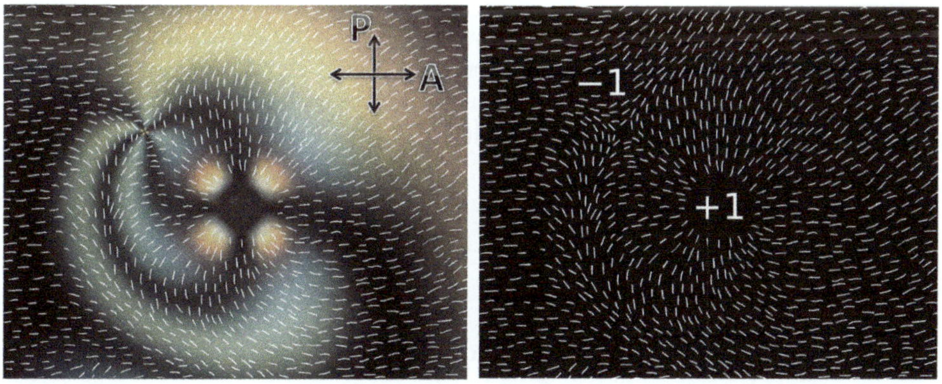

Figure 4.5. On the left is a reproduction of the micrograph from figure 4.4, this time with an illustration of the nematic director orientation overlaid. The orientations of the polarising microscope's polariser and analyser are indicated ('P' and 'A', respectively). On the right, the director structure is repeated and the positions of the +1 and −1 defects are labelled. At the centre of the +1 defect is a dark region of isotropic phase.

Figure 4.6. An illustration of the director structure around several types of liquid crystal defect. Left is a +1 defect, centre is a −1 defect, and right is a +0.5 defect.

Figure 4.7. Droplets of nematic liquid crystal forming out of isotropic.

In figures 4.3 and 4.4, the liquid crystal is held very close to its isotropic transition temperature. Round isotropic regions appear dark, and liquid crystal that surrounds these regions prefers to align with the director perpendicular to the isotropic–nematic boundary. This effectively results in a +1 defect structure around these isotropic bubbles. If a +1 defect is created, a −1 defect must also form to balance the topology, and these are visible close by. If the sample were cooled, the isotropic region would shrink to a +1 defect which would then annihilate with its neighbouring −1 defect.

4.4 Isotropic-to-nematic phase transitions

When water freezes, solid crystals form that have sharp and regular shapes. In contrast, the isotropic-to-nematic phase transition is a change from a liquid into another fluid. Rather than crystals, we get birefringent droplets that make for beautiful displays of colour under a polarising microscope, as in images 4.7 and 4.8.

Figure 4.8. Nematic liquid crystal forming out of the isotropic phase.

To produce the thin layers of nematic photographed so far, a puddle of liquid crystal material is placed on a glass substrate, which is heated to just above the transition temperature to the isotropic phase. Cooling the surface of the puddle by blowing gently on it, a thin surface layer of the material becomes liquid crystalline, floating on the isotropic puddle underneath. Cooling further, this nematic layer becomes thicker and in some regions spans the entire puddle, interacting with the glass substrate, which is shown in figures 4.9 and 4.10.

Untreated glass tends to be dirty or uneven. The orientation of liquid crystal molecules that are in contact with the glass are influenced by this unevenness, which is then visible as a messy structure, as can be seen in figure 4.11.

4.5 Adding dispersants

We can add complexity to our liquid crystal system to produce more intricate imagery by shaking things up (literally!) and adding other types of molecules.

The nematic liquid crystals photographed here, like oil, do not dissolve in water. Adding a liquid crystal to water and shaking the mixture vigorously results in tiny droplets of liquid crystal dispersed in the water. If left to rest, the droplets will tend to coalesce back into large droplets, and separate out over time. To prevent this, a surfactant can be used as a dispersant.

Part of the dispersant molecule is highly attracted to the water, while part of it is attracted to the liquid crystal's hydrocarbon chain. It can, therefore, dissolve in the water of the mixture, and has a tendency to collect along the surface of the liquid crystal bubbles preventing them from coalescing.

Figure 4.12 shows these dispersed liquid crystal droplets, which have been sandwiched between glass substrates. The surfactant tends to align the liquid crystal

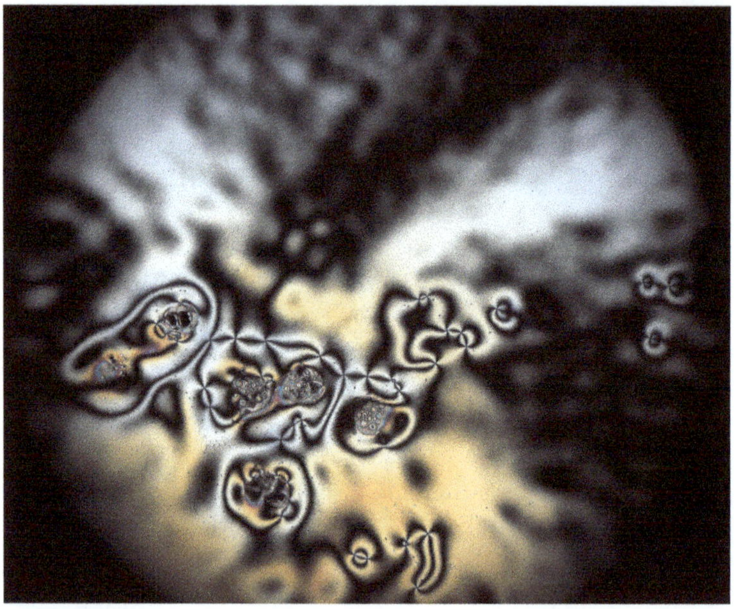

Figure 4.9. The nematic phase spreading down to the glass substrate.

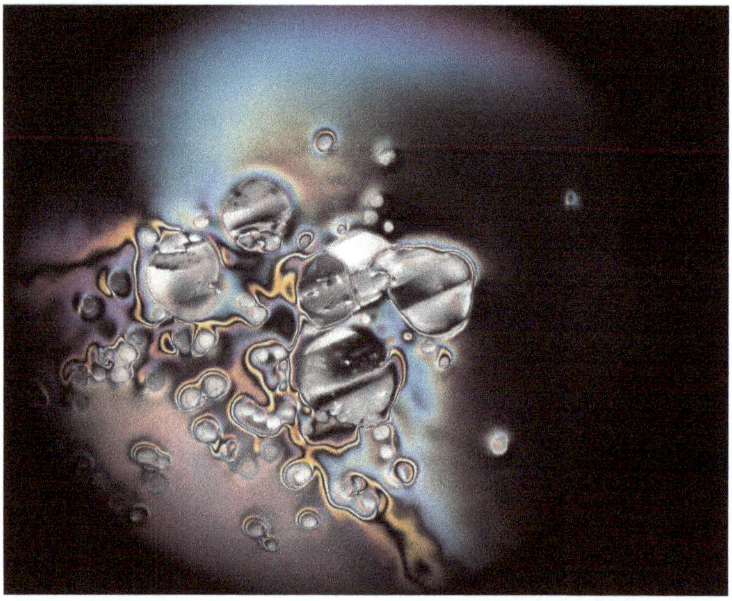

Figure 4.10. Areas in which the nematic phase spans the layer of liquid crystal material from the surface to the glass substrate form and spread over the image.

molecules perpendicular to the edge of the droplets. This sets up the director around the perimeter, but causes a topological defect to occur at the centre of each droplet. The director structures near these +1 defects have a similar structure to isotropic phase that illustrated in figure 4.6.

Figure 4.11. Nematic liquid crystal on an untreated glass substrate. The nematic director is influenced by dirt and unevenness of the substrate.

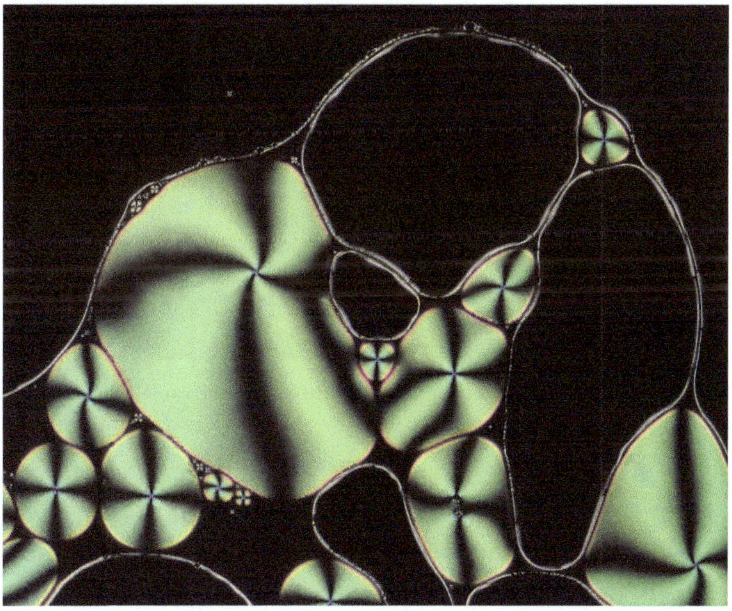

Figure 4.12. Droplets of a nematic trapped in a thin layer between glass substrates. The director alignment tends to be perpendicular to the droplet perimeter, which results in single +1 defects at the droplet centres.

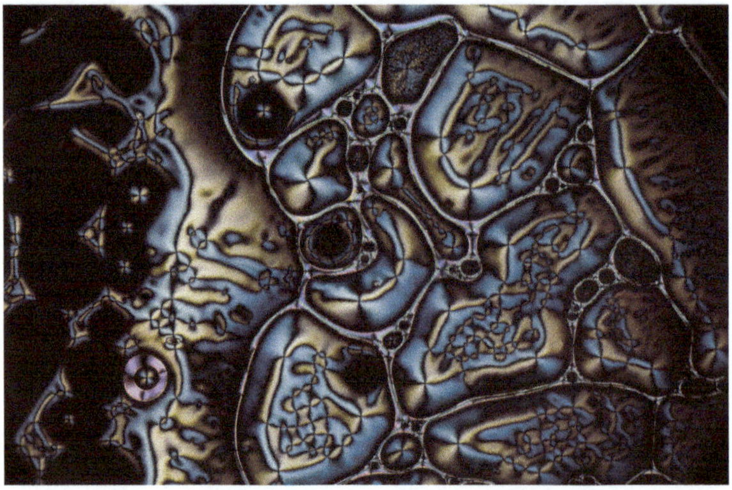

Figure 4.13. Nematic liquid crystal near the isotropic transition temperature in a preparation mixed with surfactant.

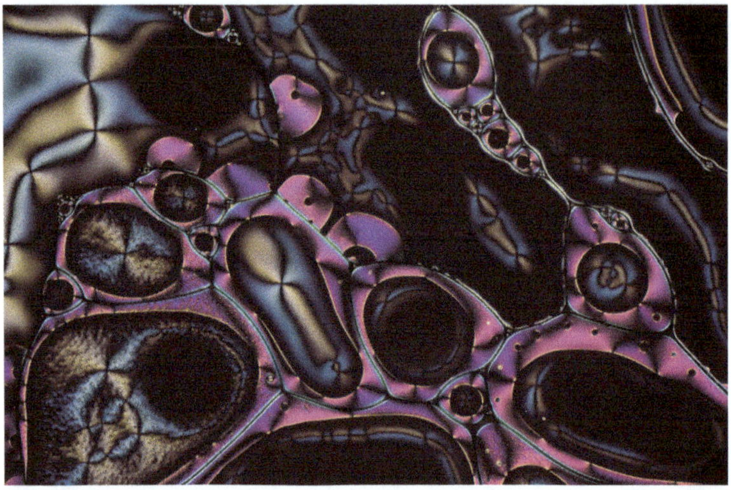

Figure 4.14. Dark regions are isotropic, while defects can be seen in the nematic phase. The mixture with a surfactant encourages the formation of structured domains, adding extra complexity to the structure.

Through a mixture of heating and cooling the liquid crystal between its nematic and isotropic phases, it is possible to create a variety of bizarre structures, some examples of which are shown in figures 4.13 and 4.14. Dark regions are isotropic, which coexist with the Schlieren defects of the nematic phase. The complex of walls separating nematic regions are the influence of the surfactant.

Liquid crystals also exhibit anisotropic flow properties. The nematic director tends to align along the flow direction, or more specifically, perpendicular to shear flow gradients. By prodding the glass substrates sandwiching the liquid crystal layer, small amounts of flow are produced. Liquid crystal droplets move and transform

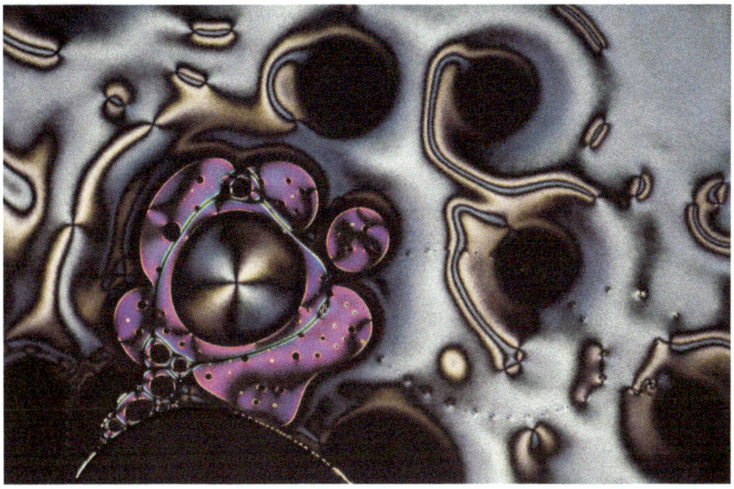

Figure 4.15. The nematic liquid crystal seen here evokes a sense of atmosphere, loneliness, and sky.

Figure 4.16. Poking the glass substrates to produce flow in the nematic liquid crystal adds to this chaotic scene.

under the flow, resulting in an added layer of complexity. This process was used to produce the images shown in figures 4.15 and 4.16.

In the next chapter, things get more complex still with the addition of non-mirror-symmetric molecules to the mix. These molecules cause the director to twist over space, resulting in bizarre and wonderful self-organising structures. But first, we will

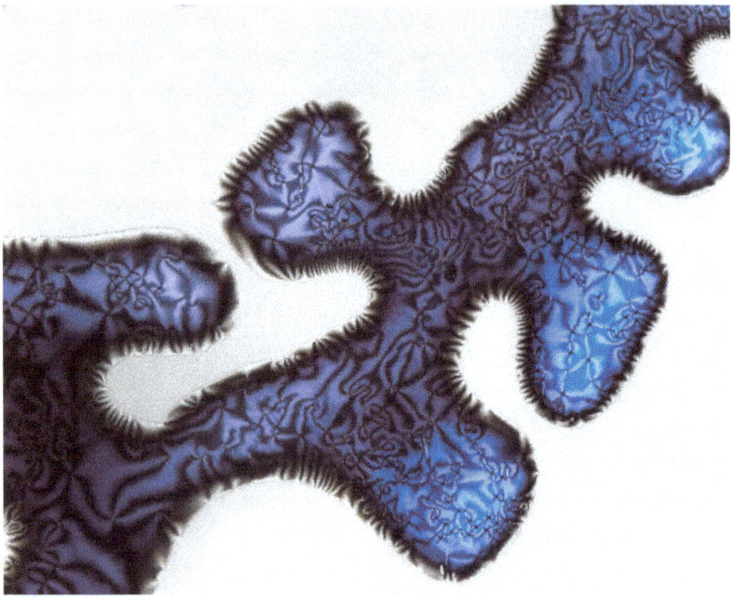

Figure 4.17. This hardened photo-sensitive glue is trapped in the Schleiren texture that it had before photopolymerisation, and is here viewed between uncrossed polarisers.

discuss how electric fields interact with nematic liquid crystals and allow us to create the most commercially valuable application (so far) for liquid crystals: liquid crystal displays.

4.6 Electric fields and liquid crystal displays

Liquid crystals are both fluid and anisotropic. We have already seen how the anisotropy interacts with the oscillating electric field of light to change the light's polarisation. Liquid crystals also interact with static electric fields.

Under a static electric field, molecules tend to rotate with their electric dipoles (the line between their negative and positively charged ends) aligned with the electric field, a bit like how a magnet will orient in a magnetic field. The fluidity of liquid crystals means molecules are free to rotate, resulting in a tendency for the nematic director to align with a field.

The liquid crystal director orientation can thus be controlled with electric fields, and in turn this can be used to control the polarisation state of the light passing though. These are the fundamentals of how a liquid crystal display works.

It's worth pointing out that there is no difference between the electric field in light and in the field used to orient the liquid crystal, except for the time scale. The electric field in light oscillates too fast for the molecules to be able to orient with it before it changes to the opposite direction, so while light is affected by the anisotropy of the liquid crystal, light does not affect the director orientation. On the other hand, if the field is static or oscillates on a time scale that the molecules have time to rotate, then it can be used to orient the liquid crystal. This is a nice example of how the same

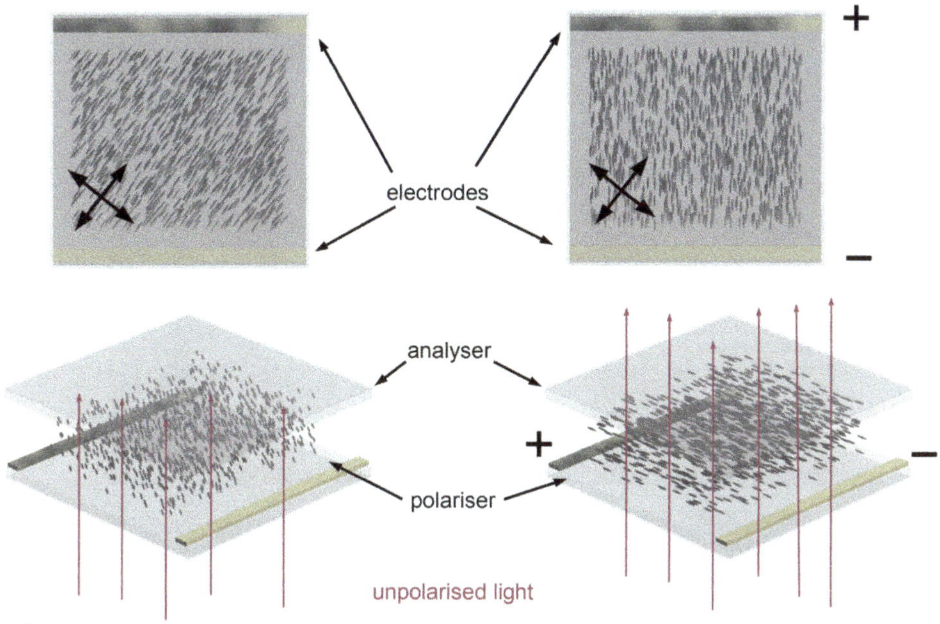

Figure 4.18. An illustration of an in-plane switching (IPS) liquid crystal display device. A liquid crystal layer is sandwiched between crossed polarisers (the polariser orientation is illustrated by the crossed arrows). An electric field, applied using rails of electrodes, is used to switch the orientation of the liquid crystal director between 0° and 45°, thereby controlling how much light can pass through the device.

simple physics can produce a multitude of different behaviours, which can then be used as building blocks for exploitation in technology, in this case to produce liquid crystal displays (LCDs).

There are a variety of ways in which the liquid crystal director orientation and driving electric field can be set up in order to produce an LCD. Many modern smartphone LCDs use what is called 'in-plane switching' or IPS LCD mode, illustrated in figure 4.18. In the IPS mode, the surfaces of the glass substrates containing the liquid crystal are treated such that the liquid crystal director is lying down in the plane of the display. Electrodes made of a transparent conducting medium (indium tin oxide) are etched into the surface of the glass such that when a voltage is applied, a field is also produced in the plane of the cell, but at 45° diagonal to the surface alignment direction. Then, polarisers are stuck to the top and bottom surfaces of the cell, crossed such that they are parallel and perpendicular to the surface alignment condition. The device is illuminated with white light from behind. The director of the nematic liquid crystal layer sandwiched between the glass substrates naturally aligns uniformly with the surface alignment, and so the device appears black between the crossed polarisers. However, when a voltage is applied to the electrodes the director orients with the field, which is diagonal to the polariser orientations. Thus, light passes through the second polariser resulting in a bright state. By carefully controlling the voltage on the electrodes, it is possible to control

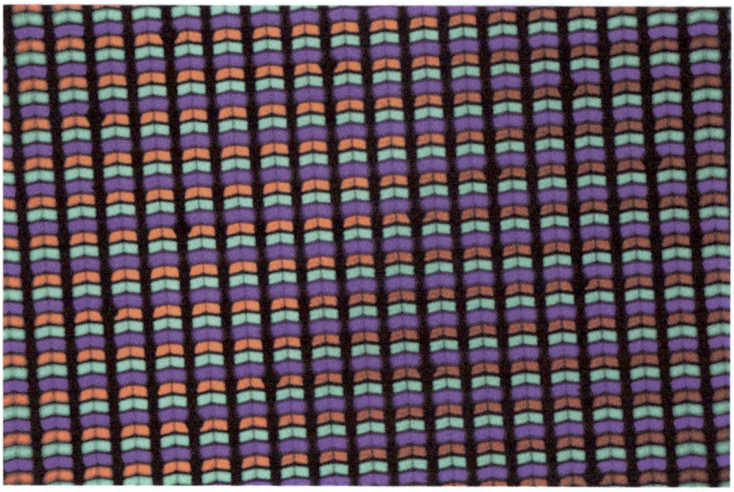

Figure 4.19. The iPhone 6 display viewed under a microscope. The red, green and blue primary colour sub-pixel structure of the display is clearly visible. These sub-pixels are further divided into regions having different orientations, which improves the appearance of the display when viewed at an angle.

how much the liquid crystal director reorients and thus how much of the light passes through.

Your eye is sensitive to three dimensions of light frequency, which we perceive as the colour spectrum. This arises directly from the biology of our eyes, which have separate light-sensitive retinal cells sensitive primarily to light at three different frequencies. Each pixel in an LCD is separated into three primary colour sub-pixels, and each sub-pixel has a filter that only allows frequencies that primarily excite one of the three types of retinal cells. Each sub-pixel is given a tiny machine, called a thin-film transistor (TFT) which helps control the precise electric field applied to that sub-pixel, and the whole lot of pixels are addressed through a matrix of thin conductive tracks that criss-cross the entire display.

The sub-pixels in modern smartphones are tens of micrometers in size, resulting in crisp displays with pixel densities in excess of 300 pixels per inch. Figure 4.19 shows the IPS LCD of an iPhone 6 under a microscope, which allows us to clearly see the sub-pixel structure. Notice that each primary colour sub-pixel is itself divided into two regions. Regions have the applied electric field and liquid crystal alignment oriented at 90° from each other, which helps to improve how the display looks from different angles.

With the advent of virtual and augmented reality displays, which are positioned much closer to the user's face, there is a push towards displays with even higher resolution. Engineers are repurposing lithographic techniques used in the manufacture of silicon transistor computer chips for producing the driving circuitry for new displays. Such techniques are routinely used to create machines that are 1000 times smaller still than existing LCD pixels.

IOP Publishing

Liquid Crystals

Benjamin Outram

Chapter 5

Long-pitch cholesterics

5.1 Symmetry

Symmetry is one of the most important concepts in science. At the forefront of physics research today is understanding how the symmetry of the Universe is broken at high energies to produce the variety of particles and forces that we see around us and detect in huge experiments such as the Large Hadron Collider at CERN. At the lower energies of condensed matter physics, phases of matter are characterised by their symmetry, for example, the rotational and translational symmetry we came across already to describe the isotropic liquid phase such as water. Symmetry is at the heart of liquid crystal science, and indeed of life itself.

The first ever liquid crystals were observed in biological systems. Between 1850 and 1890, scientists observed flowing birefringence in the outer sheaths of nerve fibres and other biological fats. In 1888 an Austrian botonist, Friedrich Reinitzer, was looking at the melting temperature of a derivative of cholesterol, cholesteryl benzoate, and discovered to his surprise that it seemed to melt not once like normal compounds, but twice. Heating the compound up, at 145.5° it melted to a liquid crystal phase, which then became isotropic at 178.5°. Despite earlier observations of similar behaviour, Friedrich Reinitzer is generally regarded as having discovered liquid crystals, probably because he was the first to promote the idea that between the first and second melting points, the material was in an altogether new phase of matter, rather than being a mixture of two phases. It was Reinitzer's collaborator, Otto Lehman, who invented the term liquid crystal. Otto Lehman had invented the first polarising microscope with a controllable heating stage, a modern version of which was used to take every photograph of liquid crystals in this book.

All life on Earth we have studied has DNA with a right-handed twist. This twist is an example of a mirror asymmetry, which is where mirror versions are not identical, like your left and right hands. Molecules that have mirror asymmetry are called 'chiral'. The mirror asymmetry in the DNA passes on to many biological molecules,

doi:10.1088/978-0-7503-1362-9ch5

including the cholesteryl benzoate in which Reinitzer first identified a liquid crystal phase.

What happens to the nematic liquid crystal phase when we introduce some mirror asymmetry? The answer is that it has an effect on the liquid crystal director. Rather than a tendency towards being uniform, the director rotates through space, as illustrated in figure 5.1. Some liquid crystals are naturally chiral, such as cholesterol derivatives and many other biological compounds, but a non-chiral nematic can be made to twist by introducing a chiral material into the mixture, called a chiral dopant. Because they were first identified in compounds related to cholesterol, chiral nematics are commonly referred to as 'cholesteric' liquid crystals, or simply cholesterics.

Figure 5.1. An illustration of the structure of the cholesteric, or chiral nematic liquid crystal. Molecules have a tendency to orient with their neighbours, but at a slight angle, which results in a helical structure. On the left, the helical structure is explicitly illustrated. Five pitch-lengths are shown, where the pitch is defined as the distance it takes for the director to rotate through a full 360°. Individual molecules are far too small to observe directly, but we can indirectly observe the helical structure by its effect on the polarisation of light. The structure is responsible for many liquid crystal phenomena, including the striped patterns of fingerprint textures seen under a polarising microscope.

The pitch length—a measure of how tightly twisted the structure is—depends on how much of the chiral dopant is added, and can be anywhere from tens of nanometers to millimetres. This compares to a strand of DNA which has a pitch of approximately 3 nm, and screw threads which are about 1 mm. Depending on the pitch length, the behaviour and appearance of the liquid crystal can be vastly different, providing an opportunity to engineer a variety of photographic compositions, and to explore some curious phenomena.

5.2 Repeating helical structures

Here, we look at chiral doped nematic liquid crystals with a relatively long pitch length of between 5 and 30 μm. This is equivalent to the thickness of a human hair, but liquid crystal structures this size are easily resolvable using polarising microscopy.

You may recall from the last chapter that a surfactant can be used to disperse nematic droplets in water. Molecules tend to align perpendicular to a surfactant interface. If glass substrates are treated with a surfactant, the director tends to align perpendicular to the glass surface in what is called a 'homeotropic' alignment condition. If we fill a cell made with homeotropic alignment with a non-chiral nematic, the director would simply align uniformly perpendicular to the glass. Between crossed polarizers, it would appear an uninteresting uniform black colour. However, in the case of a chiral nematic, the naturally twisted structure is not compatible with a uniform homeotropic surface alignment condition. A compromise is reached where the helical axis lies down parallel to the surface. This reduces the overall director distortion required as the director is perpendicular to the surface at least half of the time. However, the incompatibility of the twisting director with the homeotropic surface results in the formation of repeating line defect at every half or full pitch distance along the helix axis direction.

Figures 5.2–5.4 show cells filled with a chiral nematic. The helical axis is lying down in the plane of the cell and the director alternates between vertical and horizontal through space, resulting in the repeating pattern of dark and bright stripes. The images give the impression that the stripes are three-dimensional and sausage-like. In fact, the liquid crystal has a uniform thickness over the whole layer, but the changing twisting director orientation affects how much light, and what colour, comes through, resulting in this sausage-like appearance.

5.3 Defects and distortion

Notice the forks and endpoints in the stripes visible in figures 5.2–5.4 and later images in this section. Human fingerprints display the same kind of forks and endings, and so textures that form like this are often called fingerprint textures. In general, apart from lying down in the layer, the helical axis does not have a preference for which direction it should take, so different regions end up in different directions when the material is cooled from the isotropic phase. In the nematic case, this resulted in the pretty pinwheel-like defects in chapter 4. In the chiral nematic case it leads to these forks and endpoints.

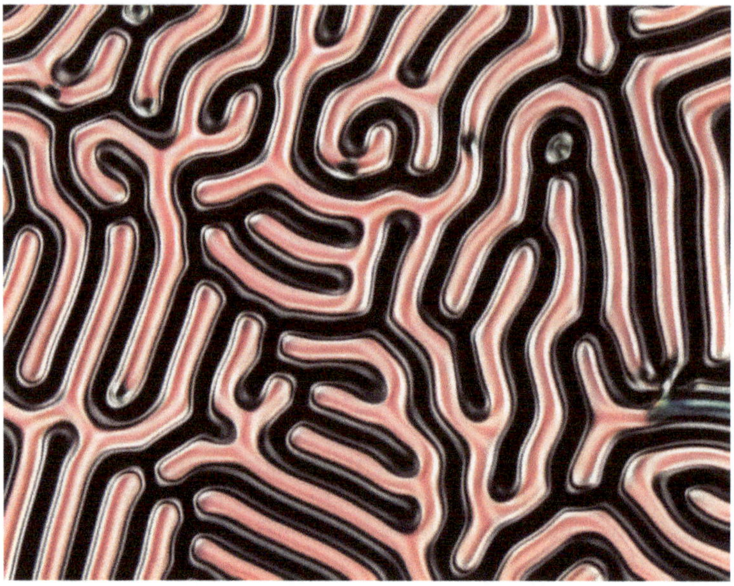

Figure 5.2. The sausage-like appearance of a long-pitch cholesteric with the helical axis in the plane of the cell.

Figure 5.3. Here, the lying helix of a long-pitch cholesteric is more-or-less uniformly oriented over the entire image. Defects in the uniform structure can be seen where the stripes fork or terminate.

Sometimes a distortion is inevitable due to the topology of the structure, for example around defects. Due to this distortion, defects themselves have a relatively large energy. In liquid crystals, there is often a compromise between the topology of the system and the relative energies of defects and distortions in the director.

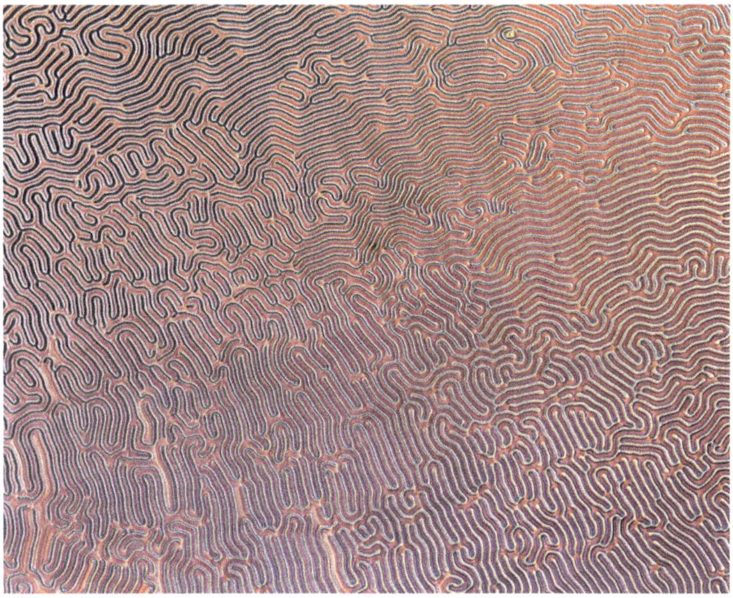

Figure 5.4. A long-pitch chiral nematic fingerprint texture. A homeotropic surface alignment condition causes the helical axis of the director to orient in the plane of the cell.

Sometimes, a small change in pitch, or cell gap, or temperature, can vastly alter the behaviour of the liquid crystal as it tends towards the local lowest energy configuration.

For example, figure 5.5 shows a fingerprint texture where there appear to be stripes of different thicknesses apparent in the same material and layer. This apparent pitch-doubling effect is due to the director becoming distorted to reduce the existence of high-energy line defects at the surface. In some parts of the cell it prefers to form line defects, and in others it prefers to distort the director. Both situations are similar in energy, which is why they appear coexisting in the image here where there is a slight pitch gradient in the liquid crystal across the cell. When we come to discuss blue phase liquid crystals, we will find even more bizarre structures that cholesterics can form due to such compromises in energy between director distortion and defect formation.

5.4 Cholesteric droplets

We can change the aesthetics of the images by putting the cholesteric in a variety of situations. In figure 5.6, for example, a thin layer of cholesteric liquid crystal has been smeared across untreated glass. This thickness gradient across the layer produces the range of colours visible under polarising microscopy. A low magnification is used in figure 5.7, which shows the corner of the glass substrates, with the long-pitch chiral nematic layer between. Finally, figure 5.8 shows a long-pitch chiral nematic being slowly melted into the isotropic phase.

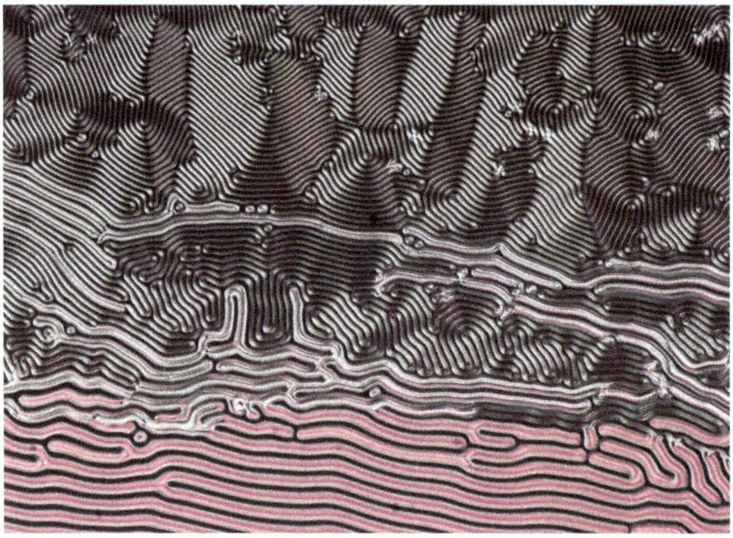

Figure 5.5. A long-pitch chiral nematic in a cell with homeotropic alignment, between crossed polarizers. There is a slight gradient in the concentration of chiral dopant in the layer, where we see a slightly shorter pitch at the top. As the pitch gets longer at the bottom of the image, the fingers become slightly more spread. At a critical pitch length, the energy required for the director to distort to the double-pitch length structure is compensated for by the reduction by half of the high energy associated with the line defects at the surface.

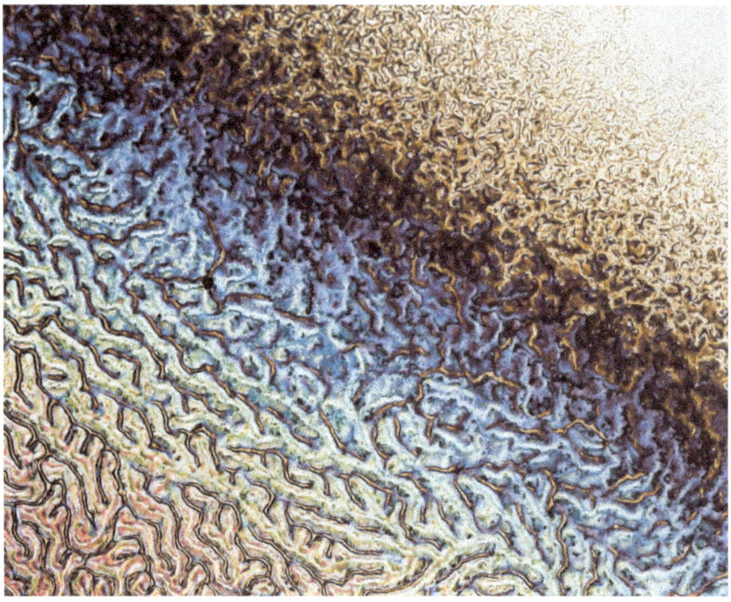

Figure 5.6. A thin layer of a long-pitch chiral nematic liquid crystal is smeared across untreated glass and viewed here between uncrossed polarizers. A thickness gradient in the layer, tapering to nothing in the top right, results in a rainbow of colour.

Figure 5.7. This low-magnification photograph, with polarizers uncrossed, shows the corner of a cell filled with a long-pitch cholesteric.

Figure 5.8. Seen through crossed polarizers, a long-pitch cholesteric is melting to the isotropic phase. The black region is isotropic.

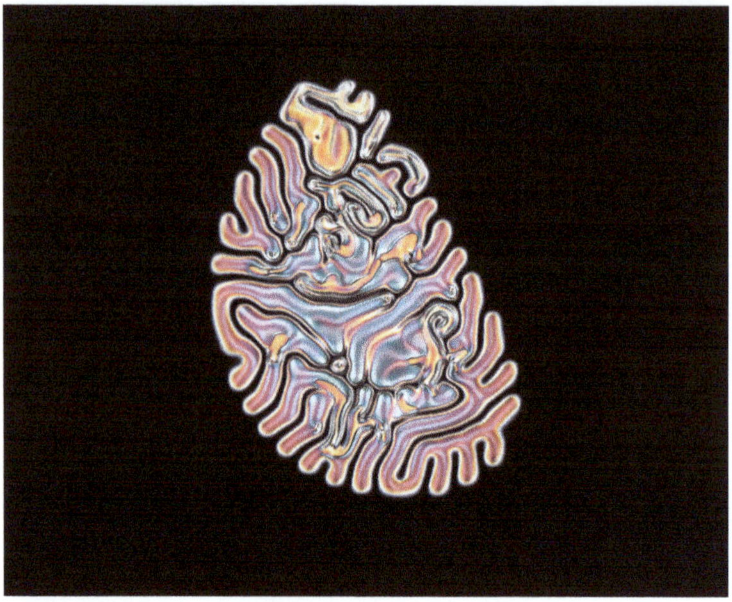

Figure 5.9. A long-pitch cholesteric droplet open to the air on glass treated to produce homeotropic alignment at the surface.

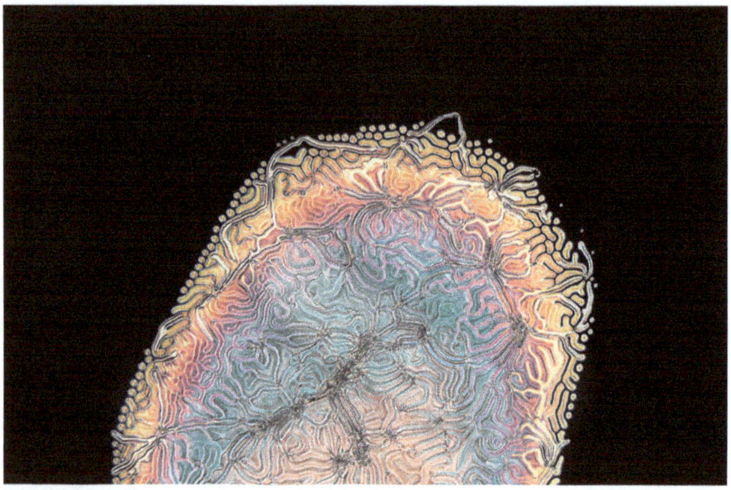

Figure 5.10. This larger droplet of the same material as shown in figure 5.9 shows a range of colour and behaviour which depends on the thickness of the liquid crystal across the droplet.

Droplets of chiral nematic on a glass surface with a homeotropic surface condition results in the behaviour in figures 5.9 and 5.10. In the first, figure 5.9, we see a speck of liquid crystal approximately 30 μm across. The thin edges show the now familiar fingerprint behaviour, which becomes more complicated as the layer gets thicker towards the centre. The second, figure 5.10, shows a larger droplet. The

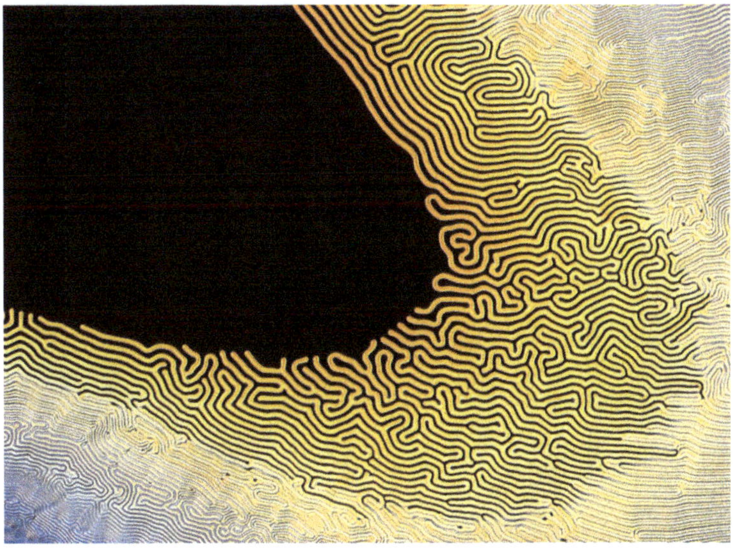

Figure 5.11. A chiral nematic liquid crystal with a strong pitch gradient. The black region is uniform homeotropic alignment due to a pitch much greater than the cell gap of approximately 10 μm. At the opposite side, the pitch is much smaller, maybe a few microns.

colours arise due to the wavelength dependence of the birefringence of the liquid crystal.

Pitch gradients can produce remarkable visual compositions. In figures 5.11 and 5.12, a non-chiral nematic and a short-pitch chiral nematic have been filled into a glass cell with homeotropic alignment, again to encourage the helical axis to be in the plane of the image. Where the two materials meet, the chiral dopant diffuses into the achiral liquid crystal, resulting in a gradient in the pitch length across the image. In figure 5.11 there is the fingerprint texture clearly visible, and the pitch length gets shorter across the image, which is visible as the striped pattern becoming more dense. The black region in this image is where the pitch is much longer than the cell thickness, and the director therefore does not have an opportunity to twist but instead remains unwound and uniformly aligned vertically, consistent with the homeotropic surface alignment. Figure 5.12 is of a different location in the same cell. Again there is a pitch gradient visible, but the polarizers have been uncrossed resulting in a bright region where the director is homeotropic, and a deep blue hue.

5.5 Out of the isotropic phase

Chiral and non-chiral nematic liquid crystal phases are identical other than for the tendency to twist over space. Phase transitions occur in the same way, albeit for the curious topologies that are generated due to the twisted structure of the chiral nematic. In some cholesterics the pitch is temperature dependent, usually increasing with temperature, especially near the isotropic transition temperature.

In figures 5.13–5.16 we see the chiral nematic phase emerging from the isotropic phase. There are the precursors to fingerprint textures forming, and droplets

Figure 5.12. Polarisers are uncrossed slightly, resulting in a different wavelength coming through the microscope analyser, resulting in different colour when compared to figure 5.11, although it is the same cell.

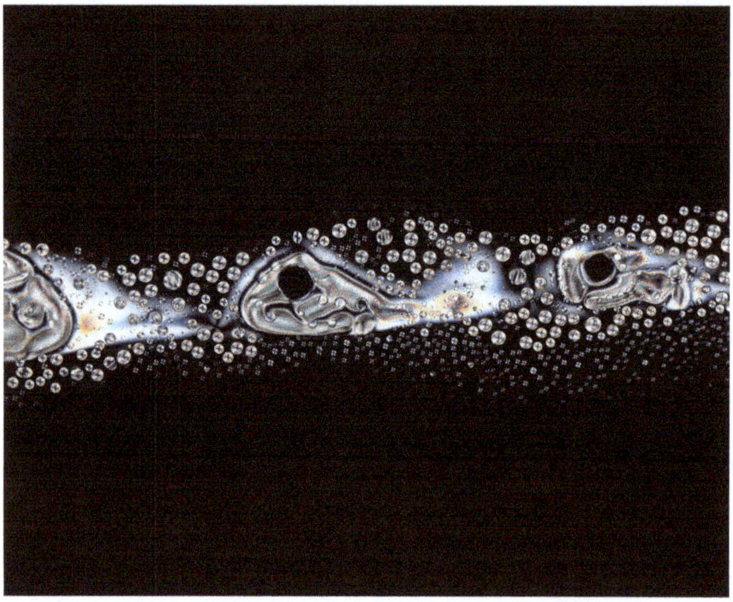

Figure 5.13. A weakly chiral nematic being cooled out of its isotropic phase. The isotropic phase is the black region.

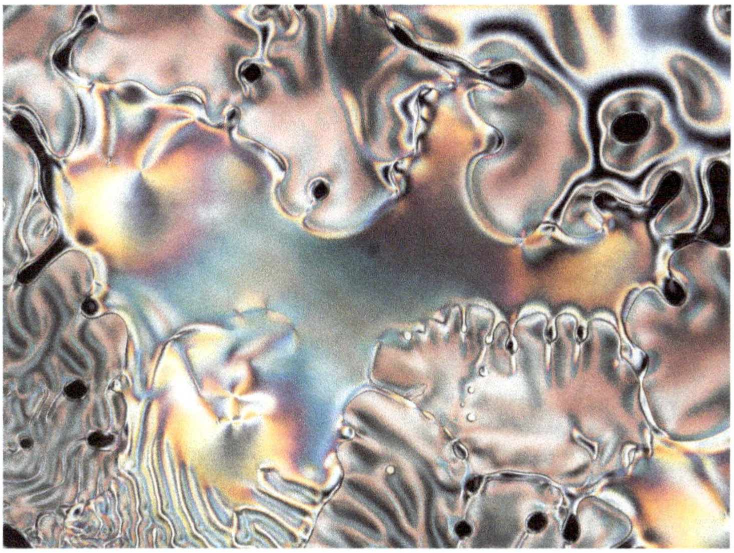

Figure 5.14. Here a weakly chiral nematic has formed recently by cooling from the isotropic phase.

Figure 5.15. Weakly chiral nematic droplets arising on cooling from the isotropic. On further cooling these droplets will coalesce and form a fingerprint texture.

reminiscent of the droplets that we saw forming out of the isotropic phase in non-chiral nematics, but with some added complexity in the structure that is thanks to the material's tendency to twist.

In figure 5.17, again it is very close to the melting temperature, and small isotropic bubbles are visible. There is a pitch gradient from a relatively tightly wound helical

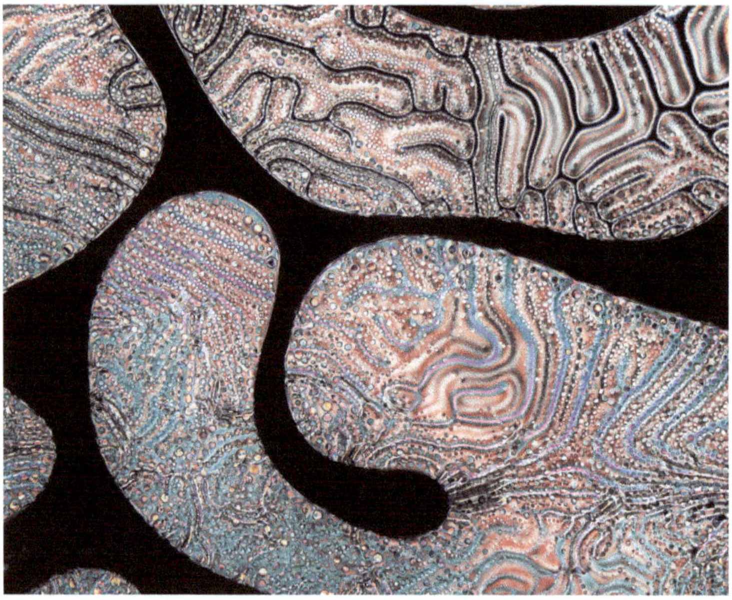

Figure 5.16. A chiral nematic forming on cooling from the isotropic phase. Black regions are still isotropic. Droplets are visible but fingerprint features have also already formed.

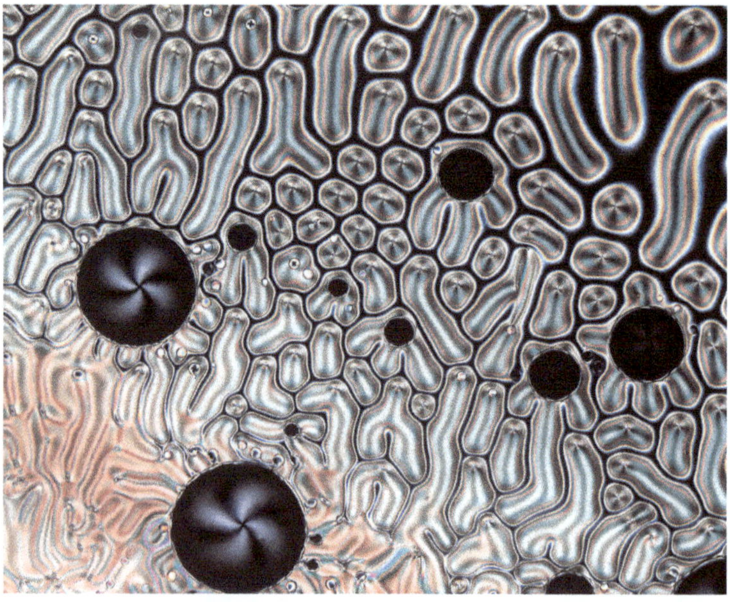

Figure 5.17. A long-pitch chiral nematic is here close to the isotropic phase transition temperature. There is a small gradient in the pitch which is slightly longer in the top right. Dark regions are isotropic bubbles.

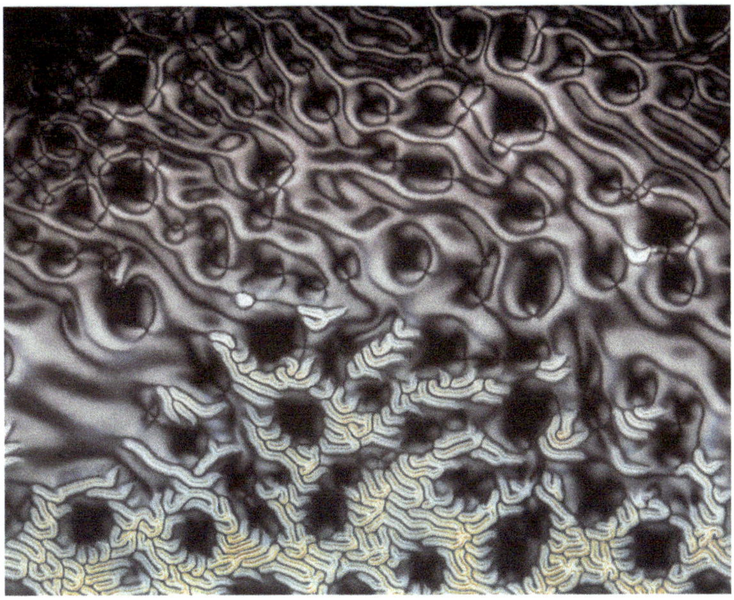

Figure 5.18. A thin skin of a long-pitch chiral nematic floating on the isotropic phase. Above a certain thickness threshold, the director tends to twist showing a fingerprint texture. Below this threshold it does not, and a Schleiren texture is visible.

structure in the pink region, with more sparse finger structures in the top right indicating a longer pitch.

Figure 5.18 shows a skin of weakly chiral liquid crystal open to the air, floating on isotropic liquid beneath. This skin is so thin that the molecules do not form helical structures. However, where the skin becomes thicker at the bottom of the image, the fingerprint texture is visible. There seems to be a threshold thickness above which the director is able to twist up.

5.6 Helix unwinding

So far we have looked at a regime in which the chiral nematic pitch is about the same as or less than the thickness of the cell it is in. But what happens if we have an even longer chiral pitch? The longer the pitch is, the more the behaviour appears like a non-chiral nematic. Above a certain threshold pitch length that depends on the thickness of the cell, a homeotropic surface alignment will completely unwind the helix and the director will align uniform and perpendicular to the glass substrates just like a non-chiral nematic, appearing black between crossed polarisers. However, if the pitch is close to this unwinding threshold, it is possible to find a mixture of some areas showing a uniform homeotropic alignment, and others in which there is a twist. This is exactly what we see in figures 5.19 and 5.20. The dark regions are where the helix is completely unwound due to the homeotropic surface alignment. However, we see bright lines over the image. These lines are 'twist walls' where the director undergoes a 180° twist across the wall. Figure 5.21 further shows such a

Figure 5.19. A chiral nematic whose pitch is longer than the cell gap. Lines represent twists in the director, and black regions are where the director is unwound and uniformly perpendicular to the glass.

Figure 5.20. Here, walls of 180° twists in the director are revealed as lines that trace out evocative shapes. The cell is treated to promote homeotropic alignment, and black regions are where the director is perpendicular to the glass substrates.

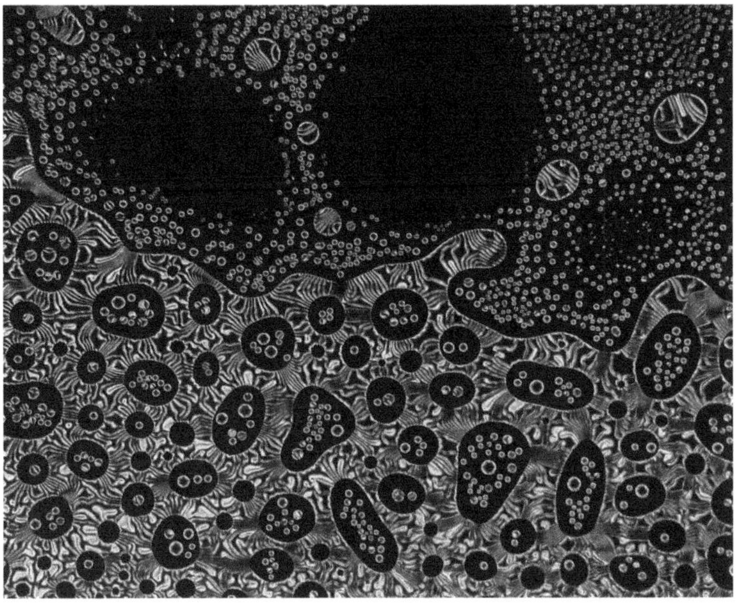

Figure 5.21. A material whose pitch is greater than the cell thickness undergoing a phase transition from isotropic liquid to chiral nematic liquid crystal.

material developing out of the isotropic phase. Some of the black regions are isotropic, and others are again where the director is unwound and vertical.

We came across how a static electric field interacts with liquid crystals in section 4.6. There, it was used to control the orientation of the liquid crystal director to allow for the creation of a liquid crystal display. Here again we are using the field to influence the director orientation behaviour. In figures 5.22–5.24, the surfaces of the cell have been treated with a thin layer of a transparent conducting medium, which allows us to apply an electric field across the device. By controlling the strength of this field it is possible to carefully tune the energy balance between the untwisted homeotropic aligned director state which appears black, and the twisted state.

In figure 5.22 a field has been applied that untwists the helix in most of the image, which appears black. The transparent electrode only goes up until the visible boundary between where the helix is unwound and where there is a clear fingerprint pattern. Without the field, the liquid crystal prefers the twisted fingerprint pattern. By reducing the field to very close to the helix unwinding threshold, some of the twist lines have been able to spread through the otherwise unwound and homeotropically aligned region to produce meandering lines.

In figure 5.23 the same cell is visible. In this case, a field much greater than the unwinding field has been suddenly switched off. In order to twist up, defects form spontaneously that allow for twist-wall structures to spread quickly across the cell. The photograph is captured at a moment before the whole image is filled with a fingerprint texture.

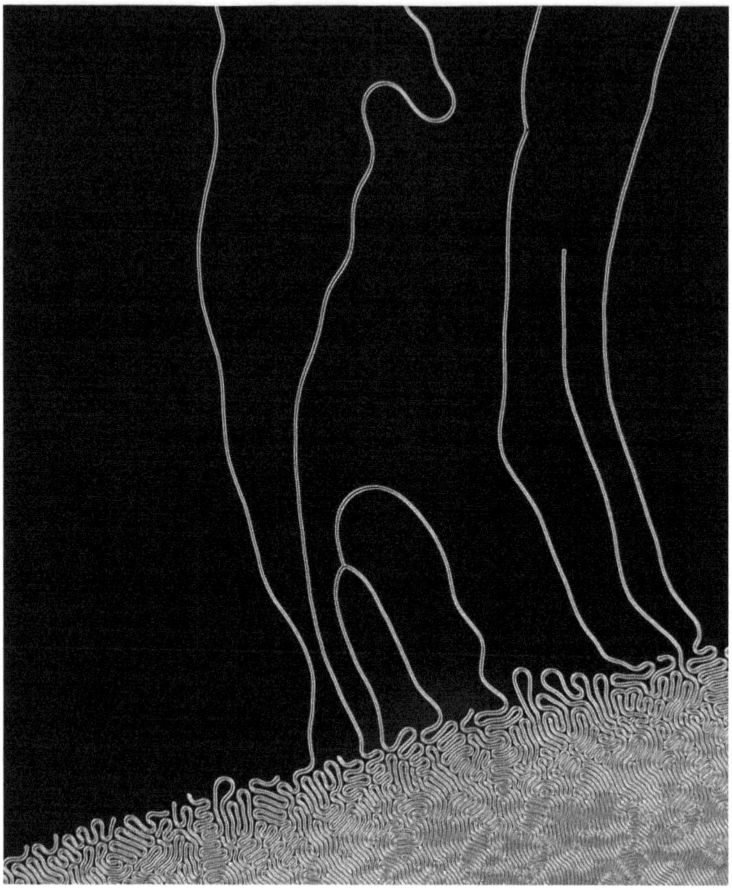

Figure 5.22. An electric field is used to tune the helix unwinding to create these spontaneous lines in which the director undergoes a half-twist. The boundary at the bottom is the edge of the electrode used to apply the field.

A different behaviour is exhibited if we reduce the field to just marginally below the unwinding field threshold. In order for the director to twist up again, it must undergo a topological change because the uniform homeotropic alignment is incompatible with a twisted structure. Defects must be created for this to happen, which costs energy. If the field is only just below the unwinding threshold, the difference in energy between the twisted and homeotropic states may not be large enough for defects to spontaneously form to allow the structure to twist up. However, if there is already a defect available somewhere, it can move through the liquid crystal and spawn the twist in the director as it goes, producing some very interesting effects. Figure 5.24 shows such a scenario. At the tip of the fern-like structure exists a liquid crystal defect. On one side of the defect is a 180° twist in the director structure, and on the other side the director is uniform and homeotropic. Since the field is kept so that the twisted configuration is slightly favoured compared to the untwisted configuration, the defect slowly moves away from its twisted side. This results in a growing twist-wall visible as a bright line under the microscope. But

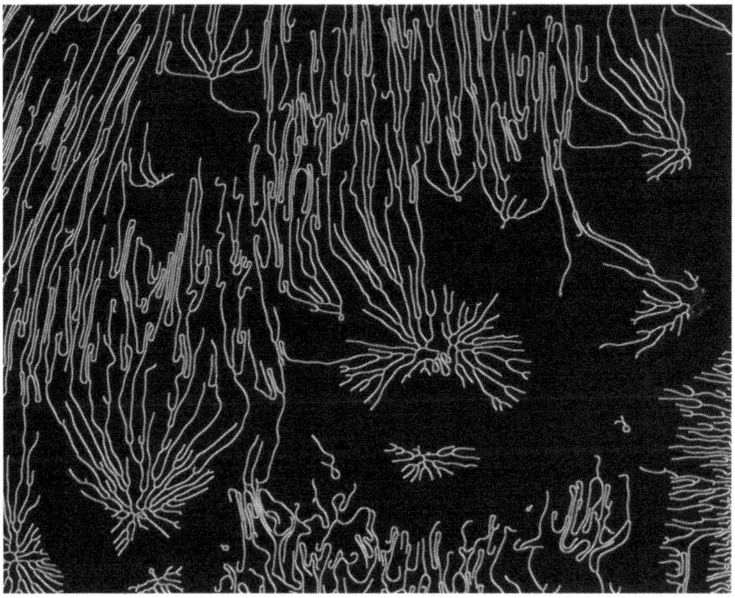

Figure 5.23. By quickly turning off an unwinding field, lines of half-twist-wall structures are seeded and spread through the otherwise homeotropic director structure.

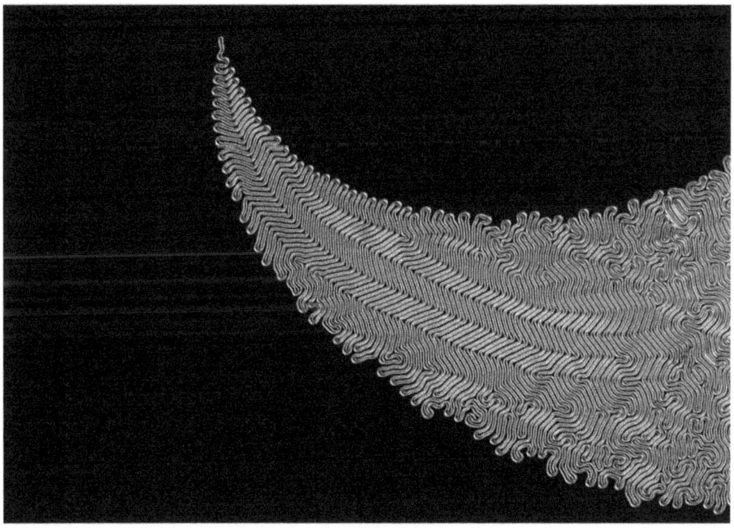

Figure 5.24. A fern-like structure results from the spreading of a single defect that seeds the formation and growth of 180° director twist lines.

it doesn't stop there; the line of twist-wall in the defect's wake itself starts to buckle and loops of wall begin to spread outwards. The difference in energy between the twisted and homeotropic structures is not enough for defects to spawn spontaneously, but the existence of a pre-existing twist-wall or defect, as we see in this example, can spread over the image to produce these fern-like structures.

The long-pitch cholesteric liquid crystals we have looked at in this chapter have been useful for illustrating their repeating helical structure, how they are influenced by homeotropic alignment, and how they interact with electric fields. They have also produced some of the most striking visual features. However, wonderful things begin to occur when the cholesteric pitch is close to the wavelength of light, or even shorter, which will be the topic of the next chapter.

IOP Publishing

Liquid Crystals

Benjamin Outram

Chapter 6

Short-pitch cholesterics

The wavelength of visible light is between 400 and 700 nm (billionths of a meter), which is about 100 times smaller than the thickness of a human hair. When it comes to how matter interacts with light, scale matters. As the size of objects approaches and becomes smaller than the wavelength of light, it becomes impossible to resolve with optical techniques, and in the case of a cholesteric, if the pitch is less than about 1000 nm it becomes impossible to see the repeating pattern of the structure with an optical microscope. But cholesteric liquid crystals can have a helical structure with a pitch of between tens of nanometers and at least several millimetres, from much shorter than the wavelength of light to much longer than the wavelength of light. In the last chapter, we explored a variety of phenomena associated with long-pitch cholesteric materials, including fingerprint textures, surface-induced alignment of the helix axis, phase transitions, and field interactions. But how do cholesteric liquid crystals interact with light when the pitch is in the range of the wavelength of visible light?

6.1 Iridescent beetles and mood rings

When light goes from one transparent material to another with a different refractive index, some of the light is reflected. This is responsible for artistic compositions in which mountains are reflected in still lakes, and why we see partial reflections from glass surfaces. Light is also reflected when it experiences changes in refractive index while moving through a liquid crystal. Liquid crystals have different refractive indexes parallel and perpendicular to their director. Imagine some circularly polarised light with a wavelength the same as a material's cholesteric pitch. The light's electric field spins around as it moves through the helical structure, and if it is going around in the same direction as the helix, it will always be aligned with just one of the refractive indices of the liquid crystal, and will pass through, no problem, like light through glass. If, on the other hand, the light's electric field is spinning around in the opposite direction, boom! It hits waves of changing refractive index as

doi:10.1088/978-0-7503-1362-9ch6

it makes its way alternatively through the refractive index perpendicular and parallel to the liquid crystal director. In this case, as with light reflected off of water, light is reflected.

With a single interface between water and air, approximately 2% of the light is reflected. In the case of light polarised in the opposite circular direction to the cholesteric helix however, it doesn't take much liquid crystal to reflect 100% of the light, as the light experiences the changing refractive index multiple times and the effect is therefore compounded. This type of reflection is known as Bragg reflection, after the father and son team who won the 1915 Nobel Prize in physics for discovering and using the phenomena to study crystal structures with x-rays.

If the range of the possible wavelengths of the light in the liquid crystal is not equal to the cholesteric pitch, then such reflection does not occur. This is because when you try to add up all the reflected electric fields, they add up to zero. The effect is that cholesterics reflect a narrow band of wavelengths that can correspond to a particular colour, and the reflected colour depends also on the angle at which you look at the material; a property known as iridescence.

Some dung beetles produce magnificent iridescent colours. The beetles have a cholesteric liquid crystal structure in their outer exocuticle layer. It is not known exactly why evolution has favoured such displays of colour, but theories include as a display for mating or as a way of reflecting heat and keeping cool. Within the same species, individuals may display a variety of different colours, and in other cases a single individual beetle may display different iridescent colours.

The pitch length of cholesteric liquid crystals is often temperature dependent. As the temperature increases, the pitch of the cholesteric may increase or decrease, depending on the material. It has been suggested that the temperature of iridescent beetles when they are in the larval stage may affect their ultimate colour, and that global warming may therefore result in the colour of beetles changing over time.

Whether or not this is true for the dung beetle, the reflected colour's dependence on temperature has been used to engineer colour-changing thermometers, battery power indicators, thermochromic paint and colour-changing mood rings. Thermochromic paint is often engineered to reflect light over the entire visible spectrum, from red at 24° to blue at 29°. This corresponds to a range between room temperature and the temperature of the human hand, so that it is highly sensitive to human touch. The Bragg reflecting properties of cholesteric liquid crystals have also been exploited for producing thin lasers whose colour can be controlled on the fly, and thin films designed to reflect infra-red for insulation.

6.2 Lying helix

As with the long-pitch cholesteric liquid crystals in the last chapter, a homeotropic surface alignment condition encourages the helical axis to lie in the plane of the cell. Figures 6.1–6.3 show a cholesteric material with a pitch of approximately one micrometer—about one fiftieth of the width of a human hair and just longer than the wavelength of visible light. At the largest magnifications possible with an optical microscope, it is just about possible to make out the repeating structure of the

Figure 6.1. A relatively short-pitch cholesteric liquid crystal held near its isotropic transition temperature in a cell with a homeotropic surface alignment condition. The helix is lying in the plane of the device, a varying helix axis orientation produces the fascinating structures visible.

Figure 6.2. The cholesteric helix is lying in the plane of the device, here seen through crossed polarizers. The eyelet structures arise from initially isotropic bubbles as the device is cooled slowly into the nematic phase.

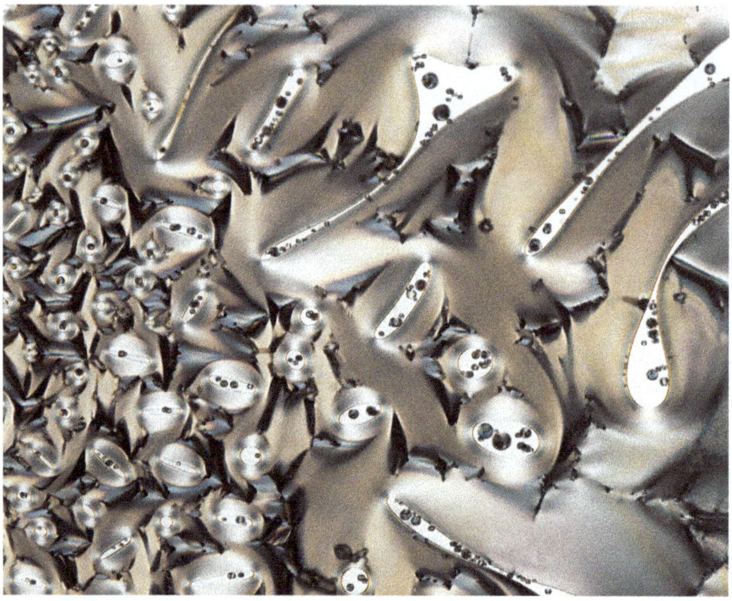

Figure 6.3. This is a different region of the same sample as in figure 6.2, however the polarizers have been uncrossed slightly which changes the dominant colour of the image. White regions are where the material is isotropic.

cholesteric pitch, which can be used to identify how the helical axis is oriented. However, it is the larger scale structures of the material that become the salient visible features. The material has been cooled from the isotropic phase and eyelet-like structures have formed around shrinking isotropic regions.

Short-pitch cholesteric liquid crystals behave in a similar way when melted or with fields applied as their longer-pitch counterparts. Figure 6.4 shows a short-pitch cholesteric forming on cooling from the isotropic phase. The polarizers are uncrossed so that remaining isotropic regions appear white. The chaotic texture contains small regions that have adopted different random orientations.

In figure 6.5 the short-pitch cholesteric has a field applied that is close to the threshold above which the field would completely unwind the twisted helical director to a uniformly homeotropic state. A field this close to the unwinding field tends to lengthen the cholesteric pitch, which results in some of the slight buckling of the otherwise straight striped patterns.

When the pitch of the cholesteric is less than the wavelength of light, less than 400 nm, the cholesteric behaves as a birefringent material with its optic axis along the helix axis (figure 6.6). The refractive index experienced by light moving along the helix axis direction is an average of the two refractive indexes of the equivalent non-chiral nematic material. Under a polarizing microscope with crossed polarizers, the material appears bright if the optic axis is diagonal to the polarizers, and dark if it is parallel or perpendicular to them. We can see this behaviour clearly in figure 6.7, where dark black regions are visible among bright oranges. The blue regions in this image are where the helical axis is vertical, which we will discuss next.

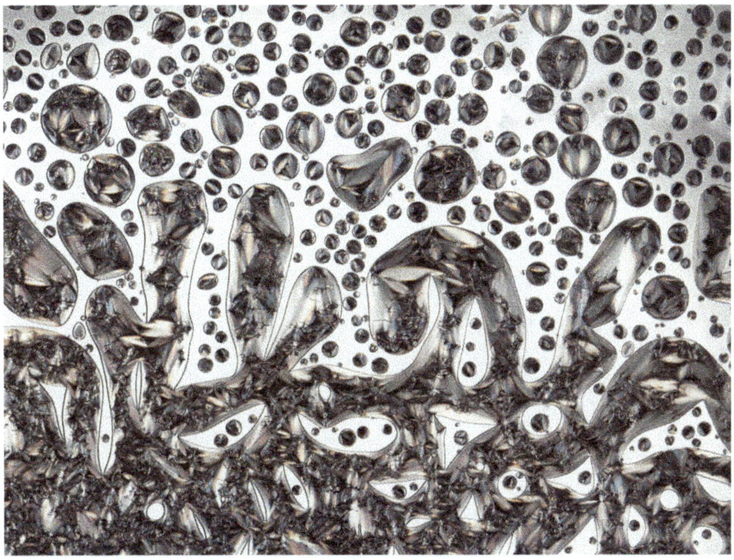

Figure 6.4. A short-pitch cholesteric forming out of the isotropic phase, here seen between uncrossed polarizers. The white regions are where the material is still isotropic. A focal-conic structure forms in the remainder.

Figure 6.5. A cholesteric in a homeotropically aligned cell with its helix axis in the plane of the device is being held here with a field close to the unwinding field applied. The field causes shifts in the pitch which buckles the otherwise uniform twist in some areas and has an affect on the colour of others.

6.3 Standing helix

So far we have looked at cases where the cholesteric helix axis is aligned in the plane of the cell, usually with the encouragement of homeotropic surface alignment

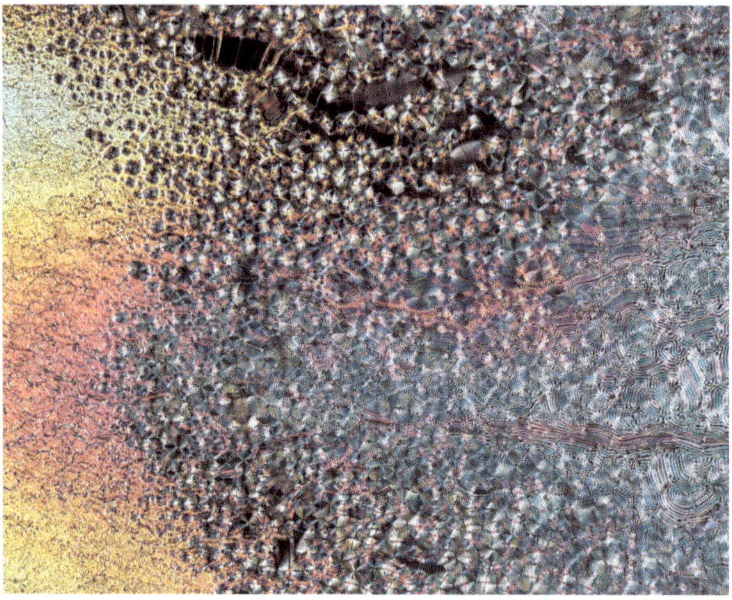

Figure 6.6. A pitch gradient across the cell from short on the left to longer on the right is visible here. The pitch length is long enough to resolve the characteristic stripey structures of the lying helix on the right. The orientation is jumbled up in many directions resulting in a rich texture.

Figure 6.7. Here we see the coexistence of regions in which the cholesteric helix axis is vertical (the blue regions) and lying horizontal (orange and pink), illustrating a sharp distinction between how light is affected depending on the orientation of the cholesteric axis.

conditions. However, the helical axis can also align vertically. A vertically aligned cholesteric helix axis is also sometimes referred to as 'Grandjean' alignment.

For a short-pitch cholesteric, the optic axis is parallel to the helix axis. Therefore, for a vertically aligned helical axis between crossed polarizers we might expect it to appear black, as would be the case with a vertically aligned non-chiral nematic. However, since the material contains mirror-asymmetry, an additional related optical effect comes into play. The light's polarization state becomes rotated through space. The degree to which it becomes rotated depends on the wavelength of the light. Shorter wavelengths are affected more than longer ones. Vertically aligned short-pitch cholesterics viewed between crossed polarizers therefore appear blue, as the blue light has undergone some rotation and been able to leak through the analyser. Referring again to figure 6.7, we can see blue regions demonstrating this behaviour where the cholesteric axis is vertically aligned, and orange and black regions where the helix axis is lying down.

Non-liquid-crystalline materials that also have mirror-asymmetric molecules, such as sugar solution, also exhibit optical activity, and also rotate the polarisation state of light.

Figures 6.8–6.12 also show regions of Grandjean (vertically aligned cholesteric). Large regions showing the characteristic blue of the Grandjean texture are permeated by streaks where the helical axis is lying down. Such a texture is known as an oily-streak texture. The exact colour the Grandjean texture appears depends on the pitch of the cholesteric, the thickness of the layer, and also on the microscope's polarizer orientations, which when adjusted result in different colours.

The Grandjean in figure 6.11 is reddish in colour due to slightly uncrossed polarizers. This cell also illustrates a couple of other interesting behaviours. The cell has a gradient in the thickness of the device with the bottom being thicker than the top, like a wedge. The glass surfaces of the cell are treated with a homogeneous

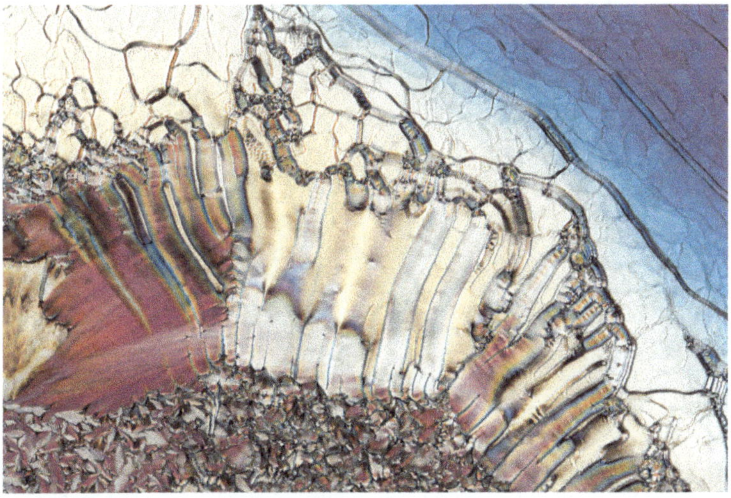

Figure 6.8. Here again we can see both in-plane and vertical orientations of the cholesteric helix. The polarizers have been uncrossed slightly which changes the colour palette for artistic effect.

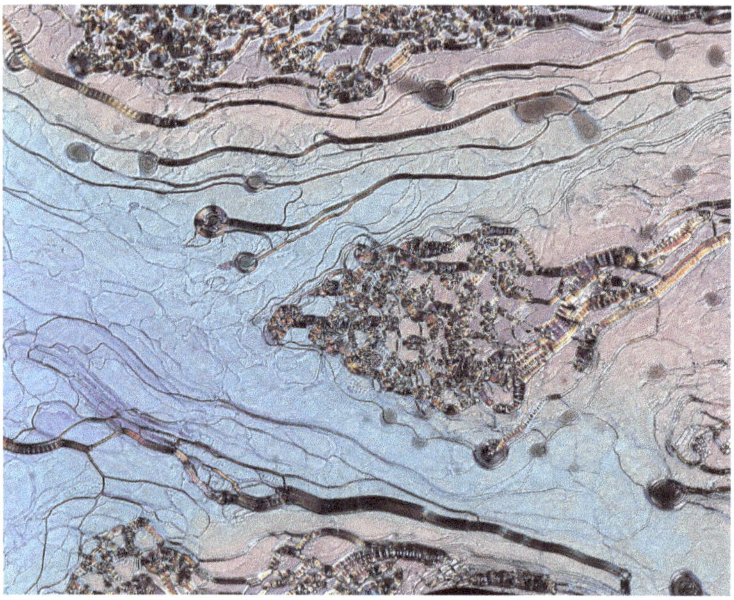

Figure 6.9. The cholesteric liquid crystal here has its helix oriented primarily in the vertical direction. There is a pitch gradient which results in the appearance of pitch jump lines, and there are line defects visible corresponding where the helix orientation is lying down. Again, the polarizers are slightly uncrossed.

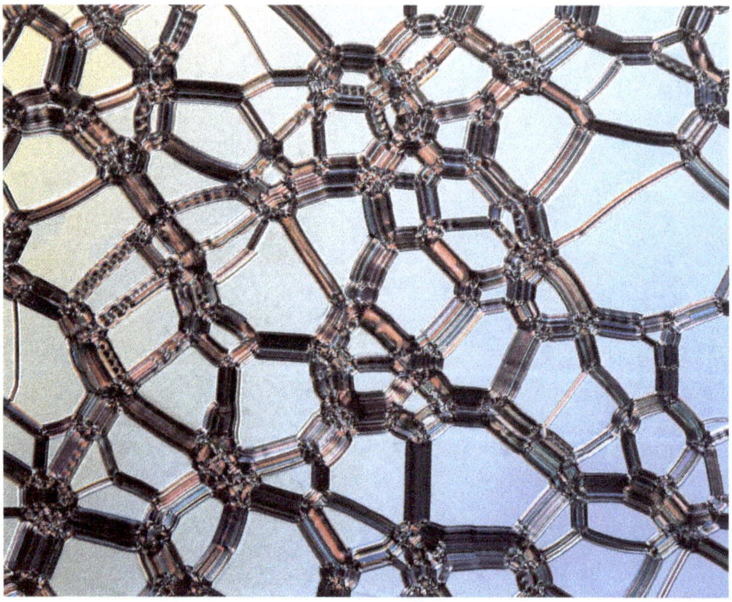

Figure 6.10. This iconic texture shows line defects in an otherwise uniformly vertically aligned cholesteric helix orientation. Yellow and blue regions are vertical, while in the colourful band structures the helix axis is in the plane and perpendicular to the bands themselves.

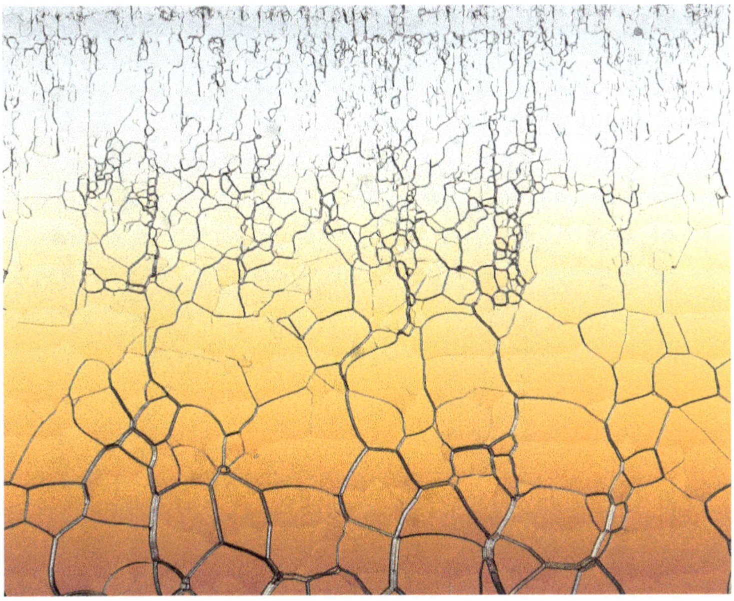

Figure 6.11. A classic short-pitch cholesteric in a Grandjean texture. A cell thickness gradient, getting thinner at the top of the image, produces these pitch jumps resulting in a terraced colour scheme.

Figure 6.12. Again, pitch jumps are visible in this Grandjean texture. Pitch jump boundary lines are punctuated by thicker defect lines in which the vertical helix alignment breaks down.

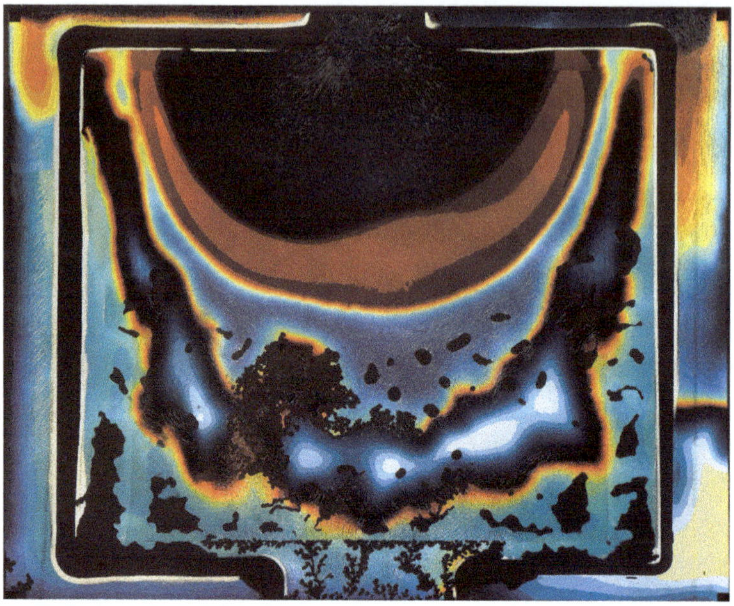

Figure 6.13. A liquid crystal cell with a uniform in-plane alignment condition is filled alternately with chiral and non-chiral nematic liquid crystal. The Grandjean structure undergoes a strong pitch gradient where the materials have been able to mix, resulting in visible bands of colour. The image is made up of hundreds of individual images digitally stitched together, and represents a width of over a centimeter.

alignment, which unlike homeotropic alignment, encourages long molecules to lie flat against the glass surface in a uniform direction. This encourages the helical axis to stand vertical, but constrains the helical structure so that it must begin and end with the director parallel to the alignment direction. The structure is therefore constrained to having a whole number of 180° twists between the top and bottom glass substrates. As the device thickness gets thinner, there are discrete jumps between regions with a different total number of rotations of the director across the layer. These pitch jumps are visible in figures 6.11 and 6.12.

The thickness of the layer and the local pitch affect the colour, so pitch jumps tend to be visible as regions of slightly different colour. The pitch can, in some cases, vary dramatically over the cell, for example if we are mixing two cholesteric materials in the same cell that have vastly different natural pitches, or if there are temperature gradients affecting the pitch. Figure 6.13 has been compiled by stitching together hundreds of polarizing microscope photographs taken of the same cell. The whole cell is visible, approximately one centimetre across. Long and short-pitch cholesteric materials have been filled into the cell (at the top of the image). By placing droplets of liquid crystal at the edge of the prefabricated cell, which has a cell gap of around 5 μm, the material soaks into the cell via capillary pressure. Long and short-pitch cholesteric materials have been filled into the cell alternately, such that bands of long and short-pitch material form in crescent-shaped regions and mix with each other resulting in arched bands where there are strong pitch gradients. These regions are visible in the image as bands of bright colours.

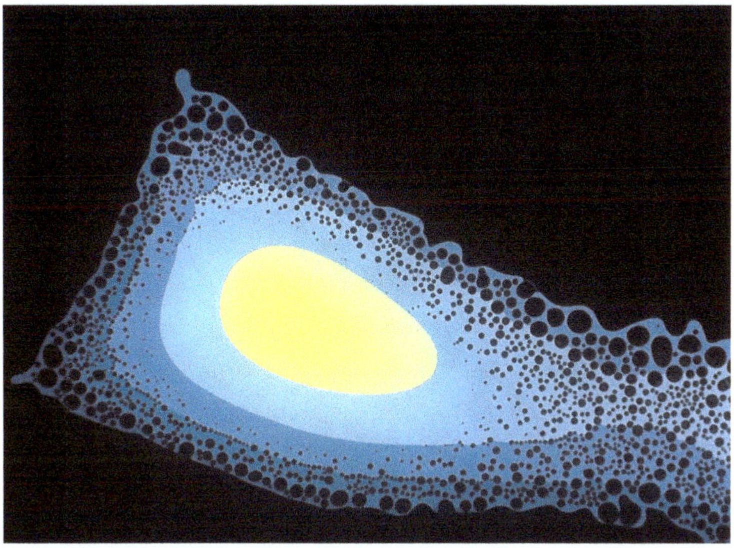

Figure 6.14. A cholesteric in the Grandjean configuration is shown here melting to the isotropic. Sharp edges to banded coloured regions result from surface alignment constraint and a temperature-dependent cholesteric pitch gradient.

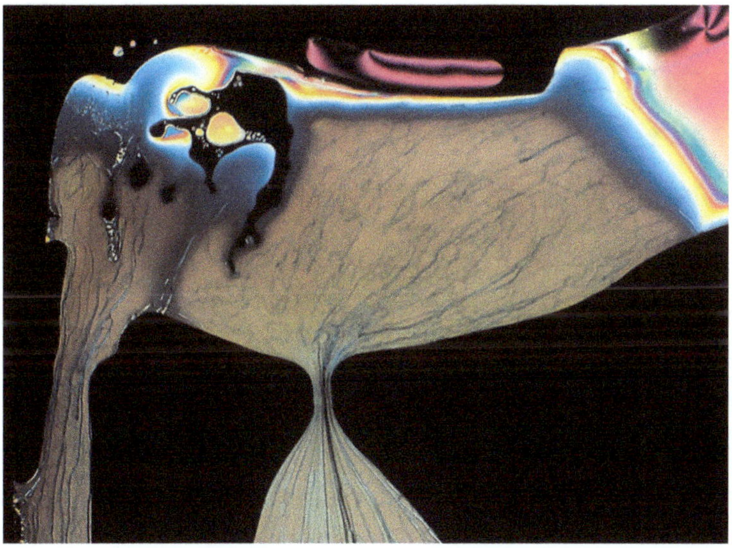

Figure 6.15. A cholesteric is mixed with non-chiral nematic in the cell, and held close to the isotropic transition temperature. Black regions are isotropic.

6.4 Phase transitions

Cholesteric liquid crystals can also melt into the isotropic liquid phase, which appears black between crossed polarisers, or crystallise if cooled. Figures 6.14 and 6.15 show cholesteric in a Grandjean structure coexisting with isotropic regions near the nematic-to-isotropic transition temperature.

Figure 6.16. A short-pitch cholesteric in the Grandjean configuration is shown here partially crystallised.

Figure 6.17. Crystals have formed along a tree-like structure here out of a cholesteric nematic phase.

Figures 6.16 and 6.17 again show the Grandjean cholesteric between crossed polarisers, this time with crystals having grown through the liquid crystal. Crystals forming from the liquid crystal material are also birefringent, resulting in a network of colours criss-crossing the texture.

Finally, liquid crystal can decompose to form gaseous phases. From a liquid crystal scientist's perspective this is usually to be avoided, but is visible in the

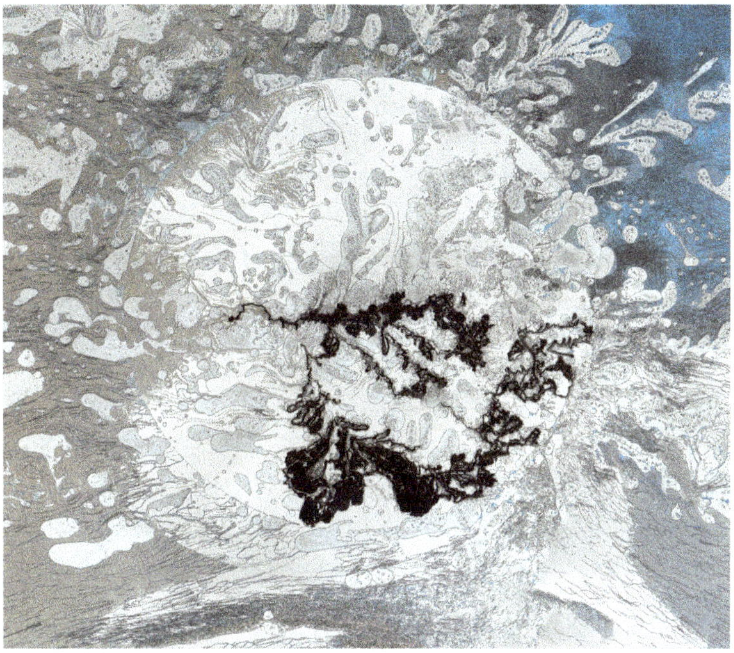

Figure 6.18. This composite image shows most of a cell of width greater than one centimeter. Polarizers are uncrossed. A long-pitch cholesteric has been subjected to an electric field via a circular electrode. The voltage applied was high enough to induce a spark, resulting in temperatures high enough for the thermal decomposition of the liquid crystal material. The decomposition products are gaseous (the white parts of the image) and solid (the dark areas).

composite image shown in figure 6.18. Here, an electric field applied across the layer was (accidentally) turned up too high resulting in a spark. This caused the decomposition of the liquid crystal into a charred black substance and a gaseous substance, both visible in the image.

Chapter 7

Blue phases, smectics and lyotropics

It was a long time before scientists reached a consensus that liquid crystals could be true anisotropic fluids. Many did not believe that a fluid would be able to have such structure, and believed them instead to be mixtures of crystals in a fluid. In fact, such mixtures also do exist and are called colloidal liquid crystals, which we will discuss briefly at the end of this chapter. However, we now know that not only can fluids be anisotropic, but they can form a multitude of structures.

As we saw in chapter 4, a nematic can be characterised by its director, which we call a uniaxial symmetry. In chapter 5 we saw that chiral nematics are further characterised by their pitch and left- or right-handedness. In this chapter we will explore the cubic symmetry of blue phases, and the layered structure of smectics.

7.1 Blue phase

Blue phase liquid crystals are a delightful example of what can come out of scientific research. Even after the discovery of liquid crystals, it is unlikely any scientist would have bet on the existence of fluids with a cubic structure. Nonetheless, such states do exist, and they may have profound technological implications.

Stable blue phases exist in some short-pitch cholesteric materials, and result from a frustrated balance between their tendency to twist and the constraints imposed by geometry. In chapter 5 we discussed how the molecules in a chiral nematic have a preference to align at a slight angle to their neighbouring molecule. As a result, the director (the local average direction of the molecules) rotates through space. As one moves along the axis direction, the director twists. But notice that the director does not twist in the direction perpendicular to the helix axis. Why not? Since molecules have a tendency to align at a slight angle to their neighbours, why don't they twist in all directions, and not just only a particular direction? In fact in some circumstances, the director can twist in two directions in what is called a double twist, as illustrated in figure 7.1.

doi:10.1088/978-0-7503-1362-9ch7

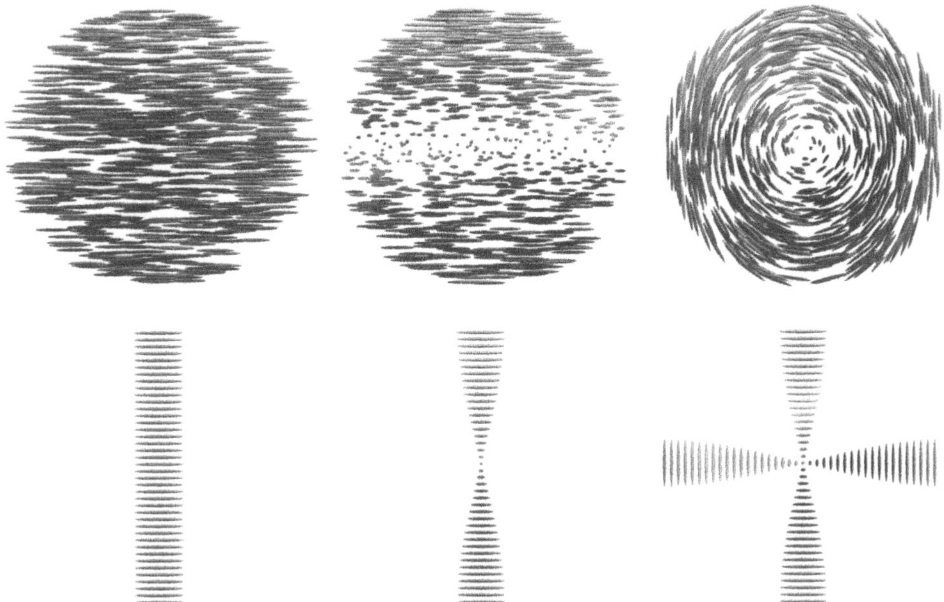

Figure 7.1. An illustration of the different kinds of twist in nematic liquid crystals. On the left is the non-chiral nematic, with its director uniform and showing no twist in any direction. The middle two images illustrate a twist in a single direction of the director, which is characteristic of the cholesteric phase. On the right, the structure twists in both horizontal and vertical directions to produce a so-called double twist (because it twists in two dimensions rather than just one). The illustration shows a cross-section through a cylinder of double twist. Such double-twist cylinders form the cubic arrays in figure 7.2, and are characteristic of the blue phases.

However, a problem arises. If the director twists radially out from one molecule, then there cannot be any twist in the direction around a circumference. The amount of twist has been increased, but only meaningfully near the centre from which the twist emanates. Away from this centre the material looks much like the standard cholesteric structure. Ideally, we would like to generate many of these double twisted structures, and pack them together closely so that more of the space is filled with doubly-twisted structure than structure that only twists in a single direction.

There is no way to pack double-twist cylinders together that does not result in defects, and defects cost energy. If the energy gained from the extra twist is not more than the energy cost of the required defects, then the double-twist structure is unstable. But if it is, then something extraordinary happens. The structure prevails and the double-twist cylinders organise into a close-packed array, forming the blue phases. Three such phases are known to exist; blue phase I, II and III. Their structures are illustrated in figure 7.2.

Blue phases tend to exist in cholesteric liquid crystals with a short pitch of the order of hundreds of nanometers, and occur as stable phases usually within a small temperature range between the isotropic phase and the cholesteric phase. While the first identified blue phases only existed over a very narrow range of less than a few degrees, scientists have now invented blue phases that are stable over a much wider temperature range, sometimes of over a hundred degrees. Such advancements are

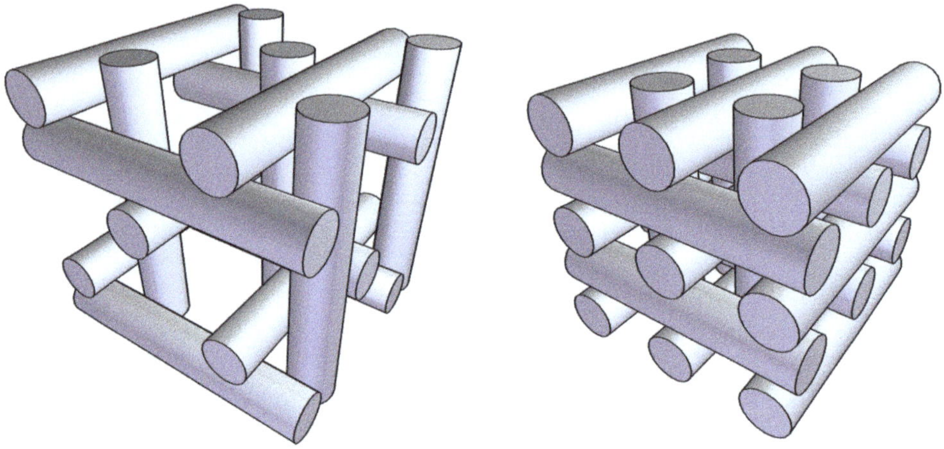

Figure 7.2. The structure of blue phase I and blue phase II are illustrated on the left and right, respectively. The liquid crystal director exhibits a twist outwards from the cylinder centres, as illustrated in figure 7.1. Defect lines (not shown) run between the cylinders.

Figure 7.3. A short-pitch blue phase liquid crystal is optically isotropic, and appears dark under a polarising microscope with crossed polarizers. Here we see a phase transition in which the cholesteric phase is branching into the dark blue phase at the bottom. The cholesteric phase grows initially with a lying helix configuration, but the characteristic blue of the Grandjean, or standing helix, is visible forming at the top of the image.

critical if they are to be used in technology. The blue phases flow like any other fluid, without losing their double-twisted structure and cubic or amorphous symmetry.

Figure 7.3 shows the blue phase between crossed polarizers, coexisting with the short-pitch cholesteric phase. Because of their cubic or amorphous structure, blue

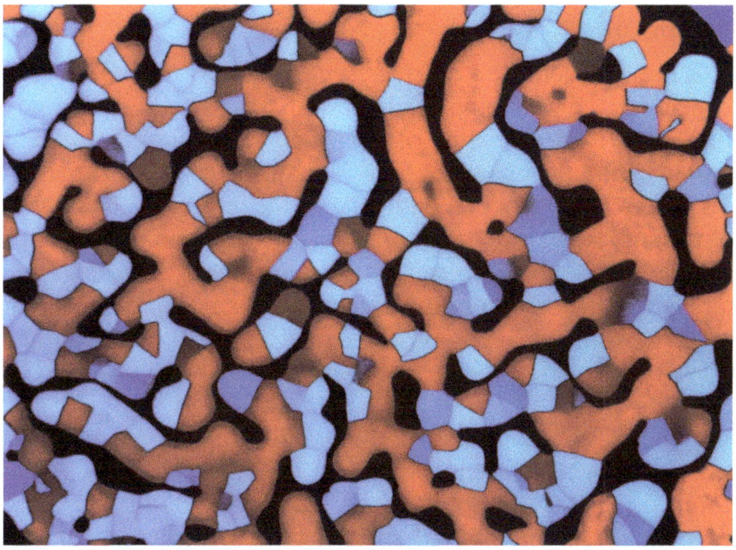

Figure 7.4. Thanks to liquid crystal photographer Christian Bohley for this image showing relatively large regions of uniform blue phase domains. The colours arise from the orientation of the blue phase structure relative to the vertical direction. Courtesy of Christian Bohley.

phases are optically isotropic—that is, they do not exhibit any birefringence at visible light frequencies. This means that under a polarising microscope, they often appear a uniform dark colour, as illustrated by the black regions of figure 7.3. Here, the material is being slowly cooled and the cholesteric phase is growing into the blue phase region.

While they are not birefringent, they do exhibit other optical properties. Their periodic structure results in the reflection of circularly polarised light that is the same handedness as the liquid crystal's twist, and that has a wavelength commensurate with the pitch length of the repeating cubic structure; what is known as Bragg reflection. Figure 7.4 shows another blue phase melting into the isotropic state. Black regions are isotropic. The coloured regions show individual domains of blue phase crystals, probably blue phase II. The pitch length is in the range of optical wavelengths and therefore light undergoes Bragg reflection. Depending on the particular orientation of the blue phase crystal lattice structure in the domains, different wavelengths, or colours of light, get reflected and result in the variety of colours that we can see, from blue to red.

Blue phases exhibit Bragg reflection from every angle, owing to their cubic structure. This makes them three-dimensional photonic crystals. Such crystals can be used to create compact lasers, and because the pitch of the blue phase can be controlled, it is possible to create a laser with variable wavelength or colour. Such devices may be used to make spectrometers that can characterise unknown materials by looking at what frequencies of light they absorb, which may in turn be used for cheap and portable medical diagnostics.

7.2 Smectic liquid crystals

Liquid crystals are characterised by their symmetry properties, as we have seen in the cases of nematics, cholesterics and blue phases. The smectics describe many liquid crystal phases, also consisting of elongated molecules, that exhibit structures with less symmetry than the nematic phase. The simplest, called the smectic A phase, has molecules ordered in layers perpendicular to the director. The smectic C phase in addition has the molecules tilted at an angle to the layers, like a book shelf with the books all slanted. Chiral molecules can be added to produce chiral smectic phases. Figure 7.5 illustrates the molecules in these smectic phases. Scientists have distinguished many other smectic phases beyond these, differing by how the layers are slanted or other physical properties. Smectic B, D, E, F, G, H, I... the list goes on.

Smectic liquid crystals are difficult to align uniformly over a large area, and this prevents them from being used in large-area displays. However, they have been heavily researched for display applications, and have found their way into small displays for projectors, cameras and specialist equipment, due to their ultra-fast switching capabilities. If the display is small, the likelihood of alignment defects is reduced to a tolerable level. For a reasonable display resolution then, the pixels must also be made extremely small. From the manufacture of computer microchips, our ability to etch small structures into silicon is highly advanced, and the same manufacturing processes can be leveraged to produce the back-plane electronics for display devices with pixels as small as 6 μm (6 millionths of a metre). Light travels through the smectic liquid crystal layer, as in the nematic liquid crystal display, but instead of passing through the device, the light is reflected off a silicon back-plane back through the birefringent liquid crystal layer. The light therefore passes through the layer twice, which means the layer can be half as thick and produce the same change in light polarisation. A thin layer also improves the switching speed of the device, which for a ferroelectric smectic display can be as fast as 50 μs (50 millionths of a second).

Figure 7.5. An illustration of the additional smectic phases exhibited by some materials with elongated particles. On the left, the crystal phase is represented with a regular array of particles. With increasing temperature from left to right, the symmetry of the phase reduces. Smectic C and smectic A phases are represented in the second and third images, respectively. As with the nematic phase (the fourth image), molecules in the smectic phase have a dominant orientation called the director. In addition, the smectic phases are characterised by their layered structure. The director of a smectic A is oriented perpendicular to the layers, and in the smectic C phase, the director is slanted relative to the layers.

Figure 7.6. A 5 μm layer of smectic A phase seen through crossed polarisers. Smectic layers have seeded out in arcs from central seed points, resulting in fan-shaped domain structures.

Figure 7.6 shows a polarising microscope image of a typical smectic phase (probably a smectic A) in a liquid crystal cell with planar alignment with a layer thickness of 5 μm. The structure forms fan-shaped domains. The director (the orientation of the molecules) is in the plane of the cell. Just like the nematic phase, this produces dark and bright regions depending on whether the director is parallel to the polariser or analyser (dark), or whether it is at 45° (bright). The colour arises due to the total birefringence across the layer, at that point. Aligned domains tend to develop out of the isotropic phase on cooling from a particular seed point, with the smectic layers forming in arcs growing outwards. This gives rise to the fan-shaped regions, which grow until reaching other growing regions. Since different regions begin growing from different points, the alignment for different fan domains is in general not the same, resulting in the complex structure you can see. The size of the domains depends strongly on how quickly the material is cooled from the isotropic phase; slower cooling results in larger domains. Such domains are also visible in figure 7.7, here seen through uncrossed polarisers, which results in different colours.

Different smectic phases are sometimes difficult to distinguish from looking merely at the polarising microscope images. Researchers use a range of other techniques, including x-ray crystallography, the application of electric fields, and differential scanning calorimetry, to identify phases and phase transitions.

You can try differential scanning calorimetry at home, using just water, a saucepan, a stove, and a thermometer. Put the water on a constant heat, measure the temperature every 20 s and plot the change in temperature over time. The temperature of the water will rise until it reaches its boiling temperature, at which

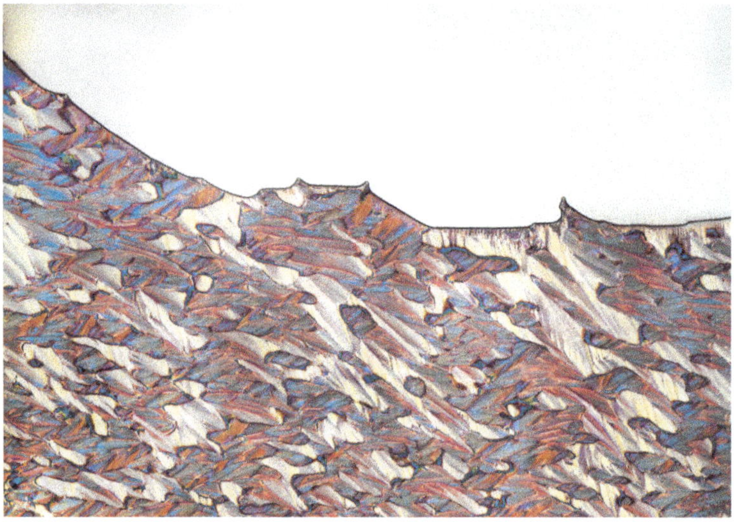

Figure 7.7. The smectic phase here borders a region of isotropic at the top of the image. The polarisers are uncrossed.

point the temperature stays at 100 °C. Once the water has all boiled away, the temperature would continue to rise. There is therefore a temporary dip in the rising temperature, and this dip reveals the existence of the phase transition. When water boils, it is obvious that there is a phase transition happening, from liquid to gas. But it is more difficult to notice changes in phase between liquid crystal phases such as smectic A and smectic C. This method of differential scanning calorimetry can be used to identify phase transitions that we might not notice under the microscope, and also to determine how much heat energy the transition requires.

The green and yellow region of figure 7.8 is probably a smectic C, seen through uncrossed polarisers. The fan-shaped domains are similar to those in the smectic A seen in figures 7.6 and 7.7, but notice the slight striation across the fans, which suggests a smectic C or other smectic structure. Also notice the strange non-coloured snake-like region that is coexistent with the smectic phase in figure 7.8. This is likely a homeotropic alignment of the smectic phase.

Smectic liquid crystals are at their most beautiful when stretched out in a free-standing film, like a bubble. Soap bubbles are suspensions of soap molecules in water, forming a lamellar lyotropic liquid crystal phase with a similar structure to a layered smectic liquid crystal. But when soap bubbles dry out, they burst. The smectic liquid crystal, on the other hand, contains only smectic liquid crystal molecules and, unlike soap, does not rely on being in a solution with water. Soap bubbles dry out because the water, which comprises small and light molecules, evaporates into the surrounding air. Smectic molecules are much larger than water molecules, and are unable to evaporate. Bubbles of smectic liquid crystal are therefore relatively robust, and can sustain for months without bursting.

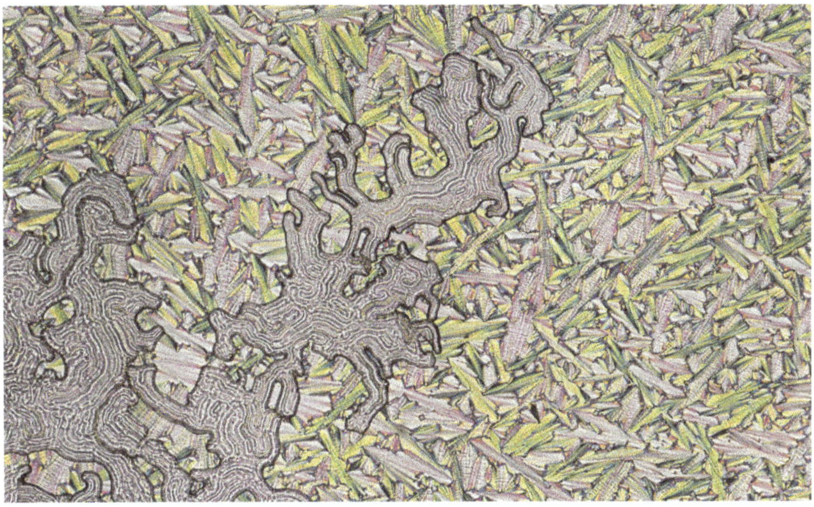

Figure 7.8. A smectic phase (probably smectic C) is shown between uncrossed polarisers. Planar aligned regions show a fan-like domain texture. The grey regions are likely homeotropically aligned regions of the same phase.

The layered structure of the smectic liquid crystal phase results in a fascinating phenomena when the material is in a free-standing film, as in figure 7.9. Molecules tend to align perpendicular to the film, with layers running parallel. The film can only be a whole number of layers thick, which leads to a terraced layer structure with different regions having different numbers of layers. Regions with different thickness appear as different colours, resulting in a cartoon-like appearance, further illustrated in figures 7.10–7.12.

Unlike the colours in most of the images in this book, the colours here arise due to the interference of particular light wavelengths reflected of the top and bottom surfaces of the smectic film. It is the same reason that soap bubbles show their iridescent changing colours as they dry and change thickness. Each step change in colour is a direct visual observation of differences in thickness of a single molecule, less than a few nanometers.

7.3 Twist grain boundary phases

The twist-grain boundary phases (TGB phases) arise in some materials at a temperature range between the cholesteric and smectic A phases. While the cholesteric phase is characterised by a strong, continuous helical structure (from chapter 5), the smectic A phase has a layered structure with molecules perpendicular to the layers. This layered structure is incompatible with a continuous rotation of the director through space, as in the helical structure of the cholesteric. The TGB phases exist as a compromise between these two incompatible structures. Figure 7.13 illustrates the structure. So-called 'grains', or layers with smectic A symmetry, are stacked next to each other at a slight angle resulting in a twisted super-structure.

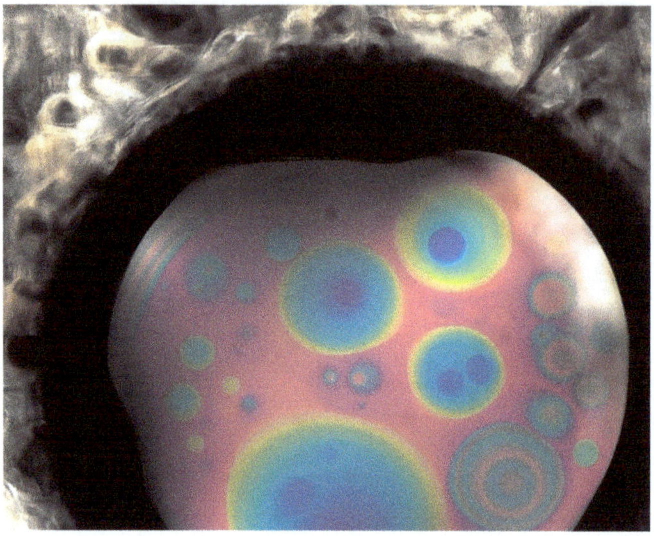

Figure 7.9. The colourful region shows a smectic liquid crystal film viewed in reflection. The circular free-standing smectic film is created by spreading the smectic liquid crystal slowly over a small hole that has been drilled into a plastic microscope slide. Different coloured regions correspond to areas with different numbers of stacked smectic layers. The colours arise from the interference of light from the surfaces of the thin film, with different thicknesses causing different colours.

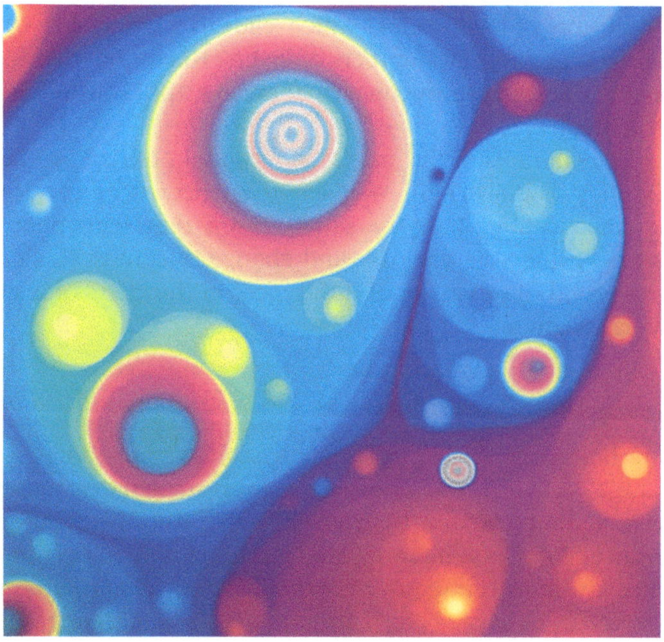

Figure 7.10. A smectic liquid crystal film viewed in reflection. Unpolarised light is shone down onto the film, reflected up through the microscope lens, and photographed. Discrete steps between regions with different numbers of smectic layers are clearly visible.

Figure 7.11. A smectic liquid crystal film viewed under a reflection microscope with unpolarised light. The black region is where the film is so thin that it does not reflect much light at all. By gently blowing on the smectic film, liquid crystal from the surrounding meniscus can be drawn into the film creating this spiral composition of different film thicknesses.

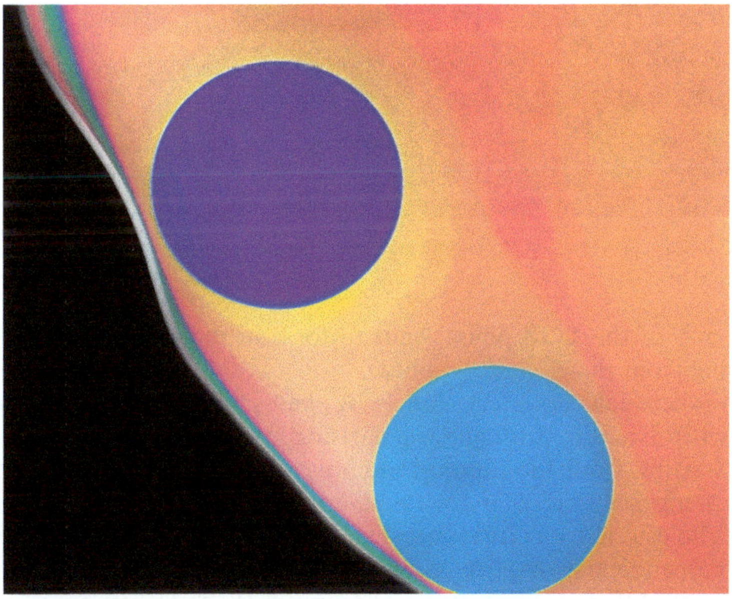

Figure 7.12. A smectic liquid crystal film under a reflection microscope. Regions of the film having a particular number of layers tend to form in circular regions. This is due to the energy associated with the step change in layer thickness, which results in an edge surface tension at the boundary of the region.

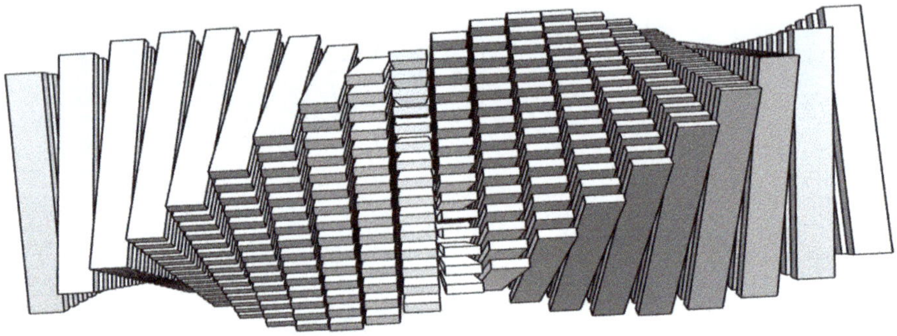

Figure 7.13. A schematic diagram of the twist-grain boundary liquid crystal phase. So-called 'grains', or slabs with a smectic A symmetry, are stacked next to each other at a slight angle resulting in a twisted super-structure. Each white block in the diagram represents a smectic A layer (see figure 7.5 to see the molecular arrangement in the smectic A structure).

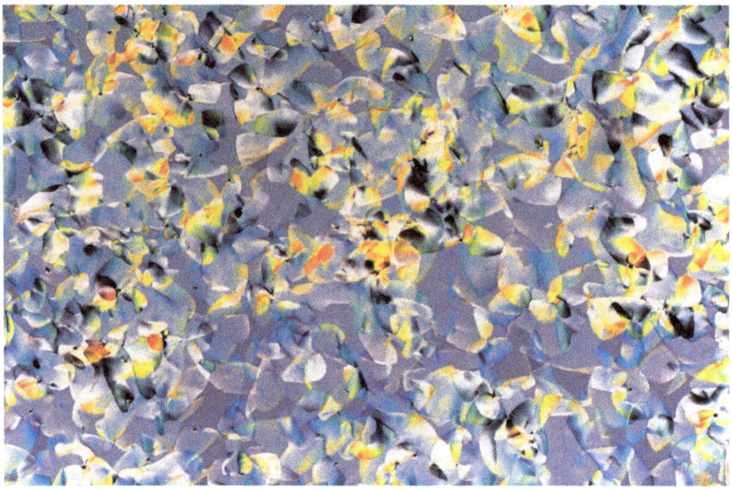

Figure 7.14. A twist-grain boundary phase growing into a cholesteric phase with a Grandjean alignment on cooling.

The helical axis of the TGB phase tends to form in the same direction that it was before cooling from the cholesteric phase.

Transitions between cholesteric and TGB phases provide dramatic floral displays of colour when viewed through a polarising microscope, as shown in figures 7.14–7.17. Starting from the Grandjean cholesteric alignment (with the helix axis of the cholesteric perpendicular to the liquid crystal layer), the material is cooled through its cholesteric to TGB phase. The helix axis of the TGB phase retains its alignment perpendicular to the layer, and different colours arise due to the development of different twist states. Figures 7.18 and 7.19 show complicated behaviour at the transition between TGB and smectic phases.

Figure 7.15. A twist-grain boundary phase, with its helical axis oriented perpendicular to the glass substrates.

Figure 7.16. Another image of a twist-grain boundary phase.

7.4 Lyotropics

Lyotropics are liquid crystals whose behaviour is chiefly governed by their concentration in a solution of water (lyo- meaning dissolve and -tropic meaning change). Their behaviour also depends on temperature, as is the case with the thermotropic liquid crystals we have seen so far. This added dimension—their concentration—allows for the level of complexity in how molecules self-assemble into higher-order structures without which biological systems and life could not exist. Common examples include soap and detergents and the phospholipid membranes of living cells. Lyotropic liquid crystals are very sensitive to changes in temperature, water concentration, and chemical environment, allowing them to have complex functionality.

Figure 7.17. Another image of a twist-grain boundary phase forming out of the cholesteric phase with Grandjean alignment.

Figure 7.18. This is likely a TGB phase forming out of a smectic phase.

The fantastic self-assembling properties of lyotropic liquid crystals are the result of the hydrophobic effect. The hydrophobic effect is the tendency for non-polar substances to stick together when in a water solvent. In chapter 2 we noted that water molecules have dipoles—positive and negative ends—that tend to flip around and allow molecules to be strongly attracted to each other. If we add an oil, which comprises molecules without a strong dipole, the oil tends to form droplets and not mix with the water. This is the hydrophobic (or water-fearing) effect in action. In lyotropic substances such as soap, molecules have a hydrophobic long part that is

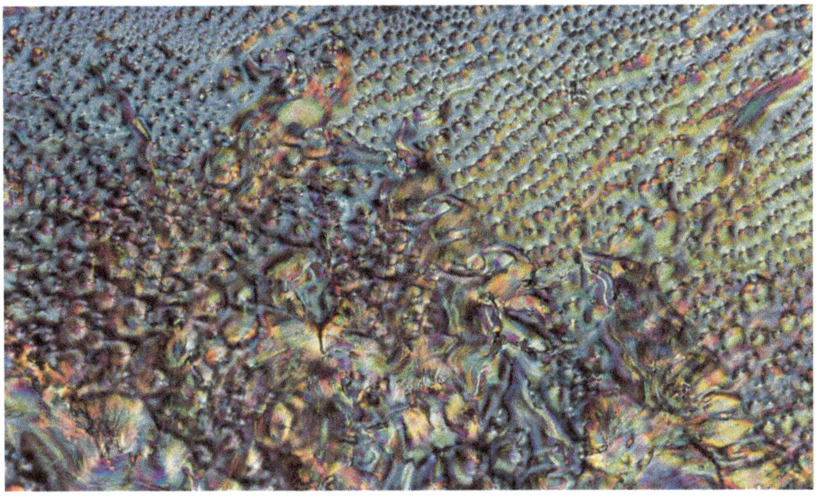

Figure 7.19. This is likely a smectic phase forming from a TGB phase, the reverse of the transition from figure 7.18.

terminated by a hydrophilic end, and such molecules are known as amphiphiles (fat-loving). The hydrophilic end is highly attracted to the water, but the hydrophobic part tends to stick to the hydrophobic part of other soap molecules. The result is that amphiphiles aggregate and self-assemble into a variety of bizarre and important structures.

When dissolved in water, amphiphiles' behaviour depends strongly on their concentration with respect to the water. Figure 7.20 shows some of the complex structures that can form. In a weak solution, the molecules are dispersed and do not exhibit any structure. In higher concentrations, the hydrophobic ends collect together, with the hydrophilic ends accumulating around the outside of spherical structures, called micelles, shielding the hydrophobic ends of the molecules from the water solvent. For the material to be classed as a liquid crystal phase, it must show a long-range ordering of molecules. This begins to happen at a concentration of approximately 25% by weight, where the spherical micelles organise into a cubic lattice structure. As the concentration increases further, the spherical micelles merge into long tubes with the amphiphilic ends of the molecules pointed towards the centre. These can self-assemble into hexagonal lattices of tubes. In even higher concentrations, bilayers of amphiphiles assemble into sheets separated by water solvent, which is called the lamellar phase.

Figures 7.22–7.24 show polarising microscope images of a detergent in water solution. The variety of stringy structures are a result of the self-assembly of the amphiphiles as they try to minimise the exposure of the hydrophobic ends of the molecules to the water solution.

The bilayered structure of the lamellar phase can form a closed bag-like structure, as is the case with mammalian cells' membranes, as illustrated in figure 7.21. Synthetic versions of these liquid crystal vesicles are being engineered to encapsulate drugs for new drug delivery mechanisms. Vesicles are made to release their drug

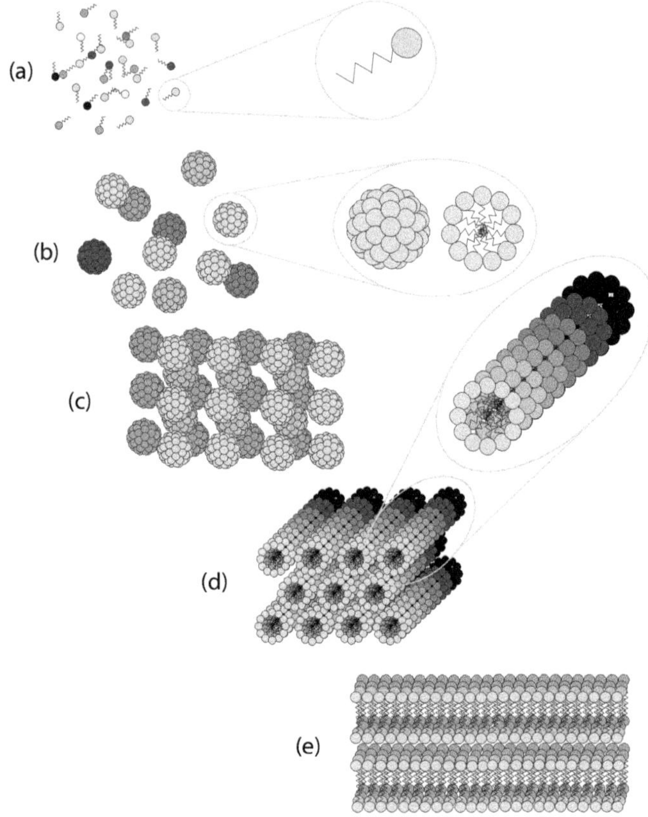

Figure 7.20. A schematic representation of the lyotropic phases. (a) Amphiphiles are molecules consisting of a water-loving polar end (represented by the head) and a fat-loving end (represented by the tail). When dissolved in water at a low concentration, they remain dispersed. With increasing amphiphile concentration in the water, they self-assemble to form: (b) dispersed micelles; (c) micelles arrange in a cubic lattice structure; (d) tube structures form and arrange in a hexagonal lattice; (e) the lamellar phase forms, in which amphiphiles form layered sheets.

payload only at the target, for example, a chemotherapy drug targeting a cancer, and thereby prevent damage from the drug to other parts of the body.

The same kind of self-assembly, due to the hydrophobic effect, governs how proteins fold, and how proteins are positioned and moved around in cell membranes. Such proteins are responsible for much of the functionality of living cells, including the cells' receptors and communications systems, transporting ions, as chemical catalysts and in the immune response of white blood cells.

Some anaesthetics and other drugs may work by affecting the liquid crystal order of the fatty membranes, which interferes with the signalling functions of cells. The lipid bilayers of nerve cells are known to be particularly sensitive to changes in chemical environment, and a liquid crystal phase transition, induced for example by the presence of the anaesthetic solvent chloroform, might affect the nerve's ability to fire, resulting in the anaesthetic effect.

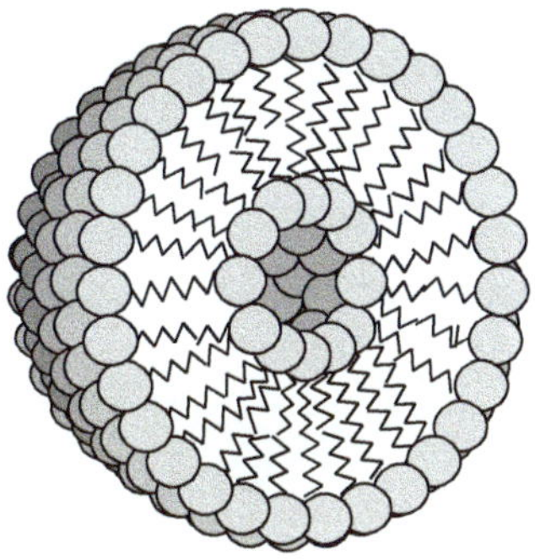

Figure 7.21. Lyotropic amphiphiles can form bilayered closed structures such as in this illustration. A cross-section is shown.

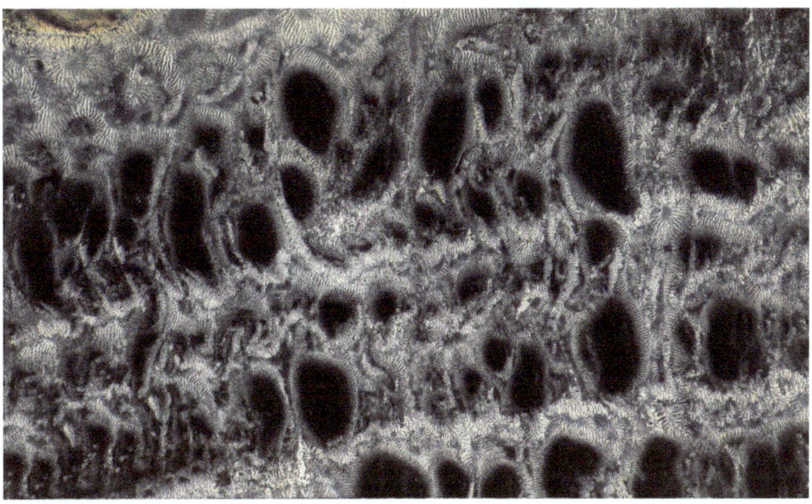

Figure 7.22. Molecules of a detergent lyotropic material have self-assembled to produce long-range structure.

It is even thought that the self-assembly of molecules via the hydrophobic effect played a vital role in the origin of life itself. Noel Clark and colleagues at the University of Boulder in Colorado showed how small fragments of DNA tend to collect together into longer chains. These chains form elongated shapes which, in solution, form a liquid crystal phase. It is thought that this self-assembly of DNA and related molecules, was vital for long enough chains to develop that would encode enough information to allow for reproduction and evolution.

There are many other examples of lyotropic examples in nature. The tobacco mosaic virus—so-called because it produces a mosaic-like discolouration of leaves of the tobacco plant—suspended in water produces a nematic-like liquid crystal phase. The virus has an elongated body like the 5CB molecule, but is orders of magnitude larger in size.

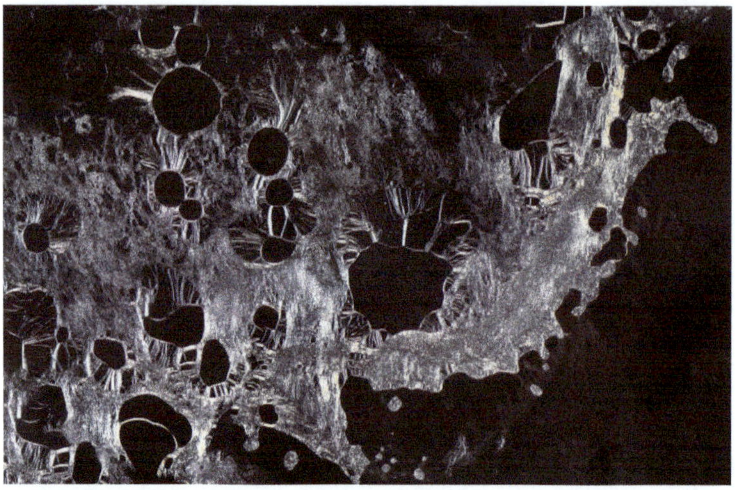

Figure 7.23. A detergent dissolved in water seen under a polarising microscope. The stringy structures have self-assembled and separated from the water solution.

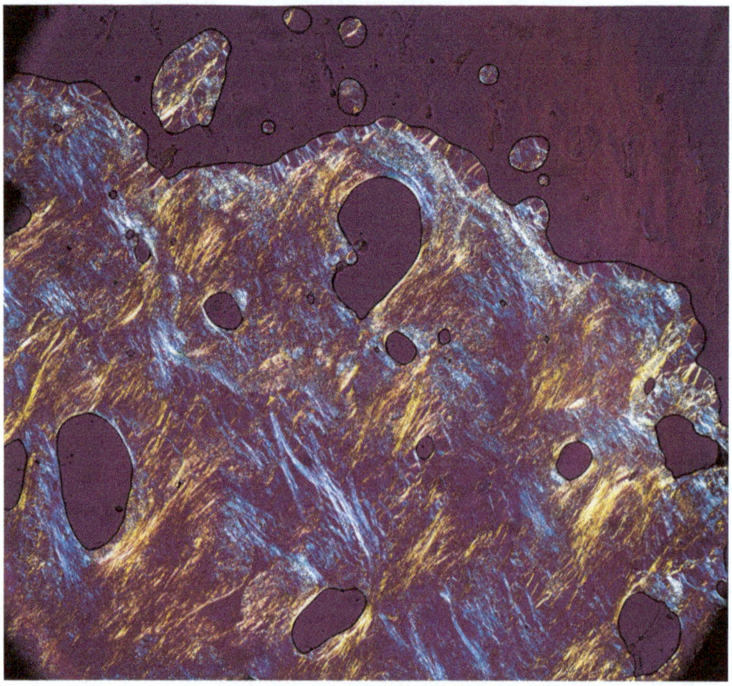

Figure 7.24. A lyotropic liquid crystal seen under a polarising microscope, in which the polarisers have been uncrossed slightly.

Natural and fairly common materials can also form liquid crystal phases when suspended in water. Sunset yellow, also known as E110, is a petroleum based orange–yellow dye used in food (such as Doritos) and cosmetics. The molecules of sunset yellow assemble into elongated structures that exhibit a variety of lyotropic

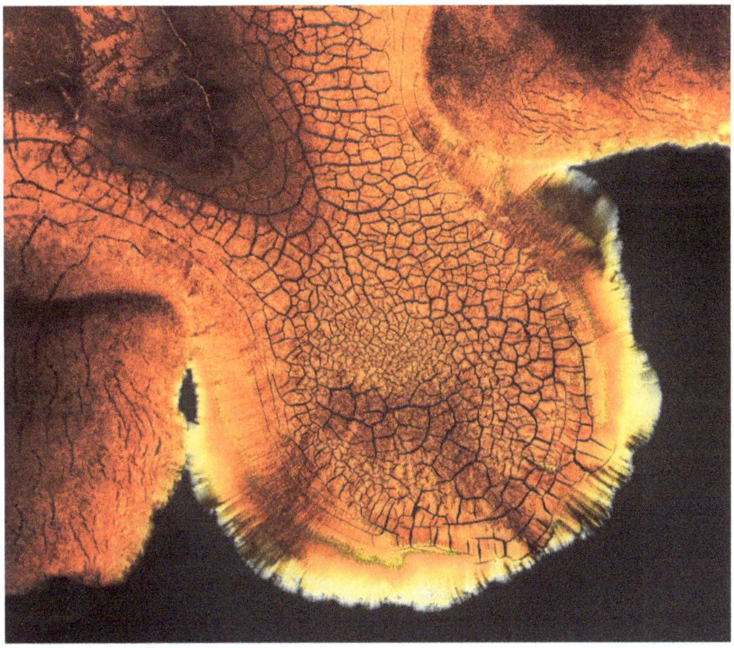

Figure 7.25. A common dye, Sunset Yellow, forms a liquid crystal phase when mixed with water in the correct ratio. Here there is a gradient in water concentration, and the coexistence of crystal and nematic liquid crystal phases.

liquid crystal phases depending on their water concentration. Figure 7.25 shows sunset yellow in water under a microscope. The material exhibits a nematic phase at the edges. As the water has dried out of the material it has left the cracked appearance in the centre. Clay particles, when suspended in water, can also result in liquid crystal ordering. Such behaviour is important to understand, as clay–water suspensions are used widely in industrial drilling, as ointments, and to make strong composite materials.

7.4.1 Spider's silk

In 2016, over 70 billion kilograms of chemical and synthetic fibres were produced worldwide. These include nylon, acrylic, polyester and Kevlar. Kevlar is used in bullet proof vests and vehicle tires, and has a tensile strength five times that of steel. Its manufacture requires high temperatures and concentrated acidic environments. Meanwhile, the lowly spider has been spinning silk for at least 400 million years. Its silk has evolved remarkable physical properties that outperform even our best synthetic fibres, despite being produced in the most unremarkable of conditions: at ambient temperatures, pressures and without extreme chemical environments. The spider can recycle its silk over and over again, while our synthetic fibres are likely to end up in landfills.

Spiders have a pouch inside them containing a liquid crystalline dope from which they spin their web silk. The dope is made from proteins, having an elongated shape, that form a lyotropic liquid crystal phase that is critical to the production of the silk. Proteins self-orient along a common director, the structure of which is carefully controlled by the water content and flow of the dope. Glands lining the spider's silk extrusion duct release a series of proteins that result in a multilayered fibre structure, with a central core, a skin, and a coating, resulting from self-assembling proteins formed from both calamitic and columnar lyotropic phases. Each layer contributes to the strength and toughness of the silk. Careful control mechanisms allow the production of different types of fibre, sometimes simultaneously, that adjust their properties based on the ambient humidity, temperature and function of the silk.

Molecules as large as the proteins in the silk dope would normally be very viscous, like honey, if it were not for their liquid crystalline behaviour. Because the molecules orient along a common director there is less viscous friction, allowing the spider's own body weight to be sufficient to draw the silk from the spider.

The liquid crystalline order results in a silk that is self-assembled down to the molecular scale. Structural defects are avoided contributing to the remarkable toughness of the web, in an evolutionary race with increasingly difficult-to-catch prey. Spiders use their silk often for multiple tasks. Depending on the spider, these may include web-building, dangling as with a rope, as a fishing rod, as a net to throw at prey, to detect movement through vibration, to build trap doors, to make structures that can be caught by the wind for transport, to mark territory and attract a mate, and to block a mate from mating with other spiders, to name just a few.

Our synthetic fibres are no match for the spider's silk-making prowess. Unfortunately, spider silk is impractical to farm due to the difficulty of keeping spiders in captivity. They tend not to get along with each other, are quite territorial and certainly not social creatures. Instead, scientists have managed to isolate the genes responsible for making silk proteins from the spiders and transplanted them into goats. The proteins are then made in the goat's milk, which can then be harvested and spun into silk.

Silk made using this process had many of the same properties as silk produced by spiders, and was produced in an aqueous (water) environment, without nasty solvents. Such silk could replace current synthetic petrochemical based fibres and plastics, especially where biodegradability, strength and stretchiness are important. In addition, spider's silk has a high biocompatibility and does not trigger an immune response, making it ideal for microsurgery.

IOP Publishing

Liquid Crystals

Benjamin Outram

Chapter 8

Discotic and columnar liquid crystals

In the last chapter, we described a mind-boggling range of structures that arise due to the self-assembly of anisotropic particles, from the cubic structure of blue phases, to the lyotropic liquid crystal phases responsible for cell membranes and cell function. Liquid crystals it turns out, are everywhere. Far from simply TV and mobile phone displays, they are vital to the function of life and are contained in cosmetics, and e-numbers. Discotic, or columnar liquid crystals add to this picture with their own unique structures and applications.

We have understood that particles that form liquid crystals must have an anisotropic shape, such as the elongated shape of the 5CB molecule, which forms the nematic phase. It turns out that disc-shaped molecules also have the shape-anisotropy required to form liquid crystal phases. Figure 8.1 shows some of the liquid crystal phases formed by disc-shaped molecules. In 1977, an Indian researcher named Sivaramakrishna Chandrasekhar discovered the first thermotropic columnar liquid crystal phases. We now know of thousands of disc-shaped molecules that exhibit columnar liquid crystal behaviour.

8.1 Discotic nematic phase

In some discotic materials, at higher temperatures the columnar structure can no longer form, but molecules still show a tendency to align with a particular direction. This is called the discotic nematic phase. It has the same symmetry as the nematic phase formed of elongated molecules (which, to avoid ambiguity, is sometimes referred to as the calamitic nematic phase). Figure 8.2 shows a messy texture of the discotic nematic phase.

8.2 Columnar phase

One popular molecule, known as HAT6, is shown in figure 8.3. Discotic materials like HAT6 tend to have rings of carbon structures forming a central unit,

doi:10.1088/978-0-7503-1362-9ch8

Figure 8.1. Phases exhibited by disk-like molecules. Disk-like molecules tend to stack on top of each other to form columns, which can aggregate into the hexagonal lattice structure illustrated in the left image. Some materials also exhibit the discotic nematic phase illustrated in the centre, which is characterised by orientational order but no stacked or hexagonal positional order. If there is enough heat energy, this nematic order gives way to the isotropic phase illustrated on the right, with no orientational order.

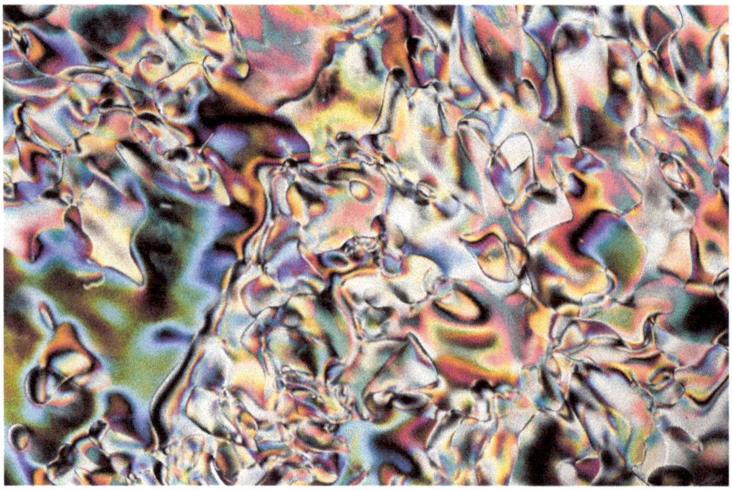

Figure 8.2. The discotic nematic phase seen here through crossed polarisers. The viscosity of the discotic nematic phase is lowest for shear flows perpendicular to the director, resulting in quite a different appearance under flow when compared with a calamitic nematic liquid crystal, which prefers shear flow parallel to the director.

surrounded by flexible hydrocarbon chains. The carbon rings tend to stack on top of each other, resulting in cylindrical columns that pack together in hexagonal or square arrays. The columnar phases are to disc-shaped molecules as the smectic phases are to elongated molecules.

In the columnar phase, electrons can move relatively freely along the columns owing to the special overlapping electron orbital structure of the carbon rings. The hydrocarbon chains, on the other hand, do not freely allow electrons to pass through

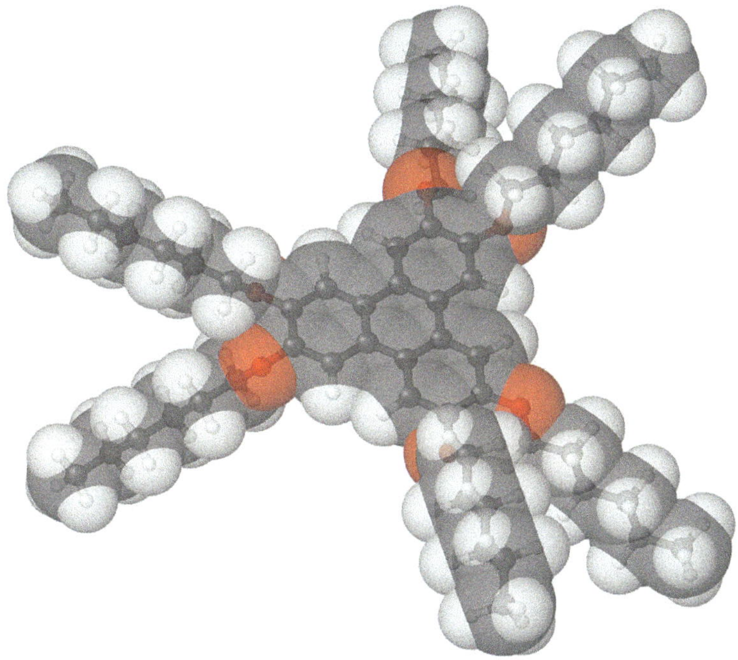

Figure 8.3. A molecule of the columnar liquid crystal material HAT6 (2,3,6,7,10,11-hexakishexy loxytriphenylene). HAT6 undergoes a phase transition from crystalline to the columnar liquid crystal phase at 62°, and becomes isotropic at 100°.

them. The result is that columns act as molecular, self-assembled, insulated electric wires. Such properties have led some researchers to explore the use of columnar liquid crystal phases in self-healing electronics, transistors and photovoltaics for solar cells.

The columnar phase can exhibit several different kinds of structures. As with nematic liquid crystals, columnar liquid crystals can exhibit defects that result in radial pin-wheel like structures. The columns of molecules are too small to see with light, but they exhibit birefringence whose effect can be seen under polarising microscopy. Figures 8.5–8.8 show examples where the columns of molecules lie on circular or spiral paths around defect centres. Figure 8.4 illustrates the structure around such defects. On cooling, multiple regions form by growing out from a defect centre until the spreading regions meet each other. You can see the straight, or sometimes slightly curved, boundaries between growth regions in the photographs. The size of the regions varies with how quickly the sample was cooled, with a slower rate of cooling responsible for the formation of larger pin-wheel-like regions.

When a columnar liquid crystal is sandwiched between two glass substrates in a liquid crystal cell, defects can begin to form on the top or bottom glass substrate surface. In this case, columns of molecules can form in a concentric cylindrical geometry out from the glass, as illustrated in figure 8.4. The second glass substrate stops further growth of the aligned columnar liquid crystal region, and what we see

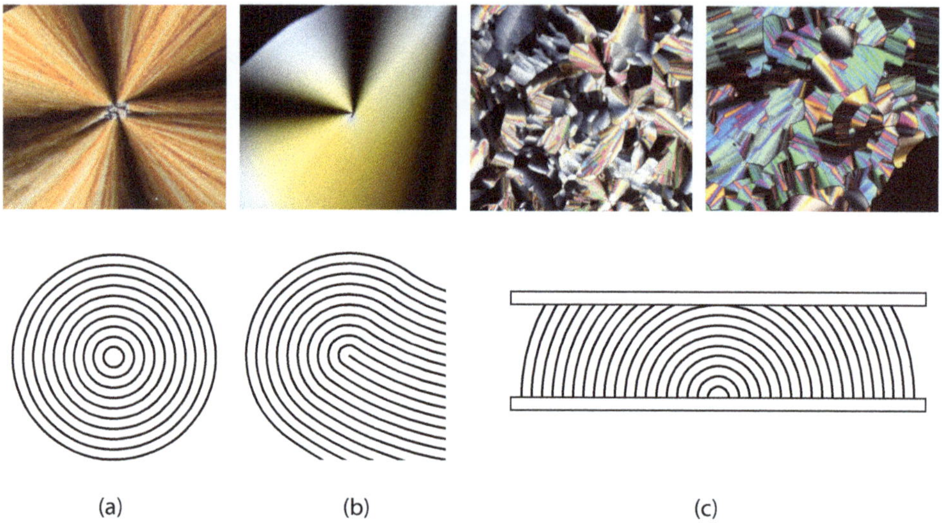

Figure 8.4. The columns of the columnar phase can fill the space in a number of orientations and topologies. In these diagrams, the lines represent the column direction. (a) Columns wrap around in a circle about a +1 defect, resulting in a 4-brush feature when viewed under a polarising microscope. Columns can also wrap around a +1 defect in a spiral structure (not illustrated). (b) A −1 defect is shown, in which columns form a semi-circular arc. (c) Columns can orient in cylindrical structures running parallel to the glass substrates (perpendicular to the viewing direction). Under polarising microscopy, these manifest as straight features that often vary strongly in birefringence as the director orientation changes across the cylindrical structure.

Figure 8.5. Here the columnar phase is seen through cross polarisers. The director (the direction along the columns) aligns in circular or spiral shapes producing the four-fold symmetric structures you see here.

Figure 8.6. A columnar liquid crystal, again forming with the columns growing in concentric circles or spirals out from defect centres.

Figure 8.7. A thin layer of a columnar liquid crystal, cooled slowly from the isotropic phase to produce large regions of oriented columns, oriented in regions around defect centres.

under the microscope is effectively a slice through the structure. Figures 8.9–8.11 show how such structures look under a polarising microscope. As we move through a slice across a cylindrical columnar liquid crystal region, the optic axis changes angle with respect to the viewing direction. As the optic axis orientation changes, so too do the colours that get through the second polariser, producing these fragmented displays of colour.

Figure 8.8. Another thin layer of slowly cooled columnar liquid crystal.

Figure 8.9. Fragmented regions of a columnar liquid crystal. The left portion is in the isotropic phase.

Similar structures can appear at the surface of an isotropic pool of HAT6 as it is cooled below the isotropic-to-columnar transition temperature by blowing gently on the sample. In figure 8.12 both vertical and horizontal concentric cylindrical column structures are visible.

Figure 8.10. A columnar liquid crystal seen between crossed polarisers. Aligned regions of columnar structure have grown out of the isotropic phase, sometimes in a cylindrical geometry. The varying angle of the optic axis in relation to the viewing direction produces different colours.

Figure 8.11. This composite image shows the intersection of two different columnar materials in the same device. Small regions are formed by columns in cylindrical structures parallel to the glass substrate (see the illustration in figure 8.4(c)).

Figure 8.12. Here, a pool of HAT6 liquid crystal has been kept close to its columnar-to-isotropic transition temperature. The surface has been blown to cool it into the columnar phase, and many fragmented regions of columnar liquid crystal have developed, with a director structure similar to that in figure 8.4(c).

Figure 8.13. A large composite image of a crystal formed below the columnar-to-crystal transition temperature.

8.3 Crystal growth

In this book we have been primarily interested in liquid crystal phases, but crystals formed by liquid crystal molecules can also be very beautiful when viewed under a microscope. Here we will look at crystals formed when columnar liquid crystals are cooled below the columnar-to-crystalline transition temperature. By controlling the

Figure 8.14. Another composite polarising microscope image of the crystal phase formed by HAT6.

Figure 8.15. A crystal phase having grown out of the columnar liquid crystal phase.

rate at which the sample is cooled, different sized structures can be created, with slower cooling resulting in larger structures. This gives an opportunity to engineer a variety of compositions by carefully controlling the rate or temperature change.

In many cases, compositions formed by slowly cooled crystals can be relatively large. Figures 8.13 and 8.14 show composite images made by stitching together dozens of photographs. The scale of these images is approximately a centimetre. Crystals have grown out of the columnar phase, spreading out from seed-points. By increasing the cooling rate, smaller crystals form that fill in the composition. The result appears like a flowery landscape. Depending on the orientation of the

Figure 8.16. Crystals are growing out of the columnar phase from the top right and bottom left of the image. Moments later, the two crystal growths met in the middle.

Figure 8.17. A crystal structure has grown out from the centre, out of the columnar liquid crystal phase, on cooling slowly below the columnar-to-crystal phase transition. The different colours are due to a different orientation of the crystal structure formation, and the resultant orientation of the optic axis.

crystal as it forms, different colours are apparent, from black and white to pink and yellow.

In figures 8.15–8.18, the crystal formation has been slowed down so as to allow very large and controlled crystal regions to develop. Crystals tend to grow from a seed point outwards, resulting in radial structures. Different colours arise due to different regions forming with the crystal's optic axis in different orientations.

Figure 8.18. Here we see well-defined crystal regions, which have grown slowly out of the columnar liquid crystal phase on cooling.

IOP Publishing

Liquid Crystals

Benjamin Outram

Chapter 9

Conclusion

From atoms and light, to liquid crystal phases and polarising microscopy, we have come quite far on our journey through nature. Liquid crystals are archetypical of the emergence of complexity from simple laws. The breaking of a simple symmetry enables a deluge of structural complexity and functionality, a principle that is exemplified by the aesthetics in the form of the polarising microscope images here, and by the success of liquid crystalline biological machines.

We saw how the anisotropy of long and thin molecules manifests at the macroscopic level in the symmetry of the nematic liquid crystal phase. By breaking the perfect symmetry of the liquid phase, we saw how the nematic can have topologies that produce flowing colour gradients and point-like pin-wheels under the microscope.

Adding just some mirror asymmetry produces the bizarre variety of forms that the cholesteric phase exhibits. Long pitches, short pitches, standing helixes and lying down helixes, birefringence and Bragg reflection, the iridescence of beetles and the thermochromicity of mood rings; all this from such a tiny change in the symmetry of the nematic phase. Not to mention the blue phases with their improbable cubic symmetry; a three-dimensional photonic crystal that self-assembles from mirror-asymmetric rod-shaped molecules.

Breaking symmetry further, the smectics and twist-grain-boundary liquid crystals with their layered architecture created mesmerising displays of colour. As they changed from one phase to another, the spontaneous reconstruction of form results in an ecstasy of the visual senses.

With the addition of amphi- and hydrophilicity, we saw how biological fatty molecules can self-assemble to form the lyotropic phases, which form the vesicles of living cells and organelles. These phases are responsible for much of the structure and function of living systems.

doi:10.1088/978-0-7503-1362-9ch9

Finally, we saw that disc-shaped molecules also have the anisotropy required to break the symmetry of the liquid phase and result in liquid crystallinity, including the discotic nematic and columnar phases.

There is, of course, a huge depth to the science of liquid crystals which this book has not been positioned to cover, and the reader is encouraged to refer to the bibliography for further reading. While the field of liquid crystal science is now very mature (this year the British Liquid Crystal Society is celebrating 50 years of liquid crystal display research), there has been a renewed interest in liquid crystal display technology for high-definition head mounted displays for use in virtual and augmented reality, including holographic displays. Other active research fields are in the use of DNA as a form of data storage, creating synthetic muscles using liquid crystal elastomers, special liquid crystal ink for anti-counterfeiting measures, understanding a variety of diseases that may be caused by the failure of certain biological liquid crystal materials, in chemotherapy and other medical treatments through the use of lyotropic membranes, in portable medical diagnostics with variable-wavelength liquid crystal lasers, for the production of new synthetic fibres such as Kevlar and those based on spider silk proteins, for compensating for distortion of light for better images of tiny biological samples, and in controlling light for optical computing. In addition, there are many exciting areas of research where there is not yet a clear application, and new liquid crystal phases are still being discovered.

Liquid crystals are a fascinating field of research, and a great meditation on nature. Complexity in the Universe unfurls from simple laws of symmetry. Grasping this concept is one of the profound pleasures in becoming aware of science, not least by witnessing it manifest in the beauty of liquid crystals. I hope you agree.

Selected bibliography

[1] Ahmad Z, Shah A, Siddiq M and Kraatz H-B 2014 Polymeric micelles as drug delivery vehicles *RSC Adv.* **4** 17028–38

[2] Bahadur B 1990 *Liquid Crystals: Applications and Uses* vol 1 (Singapore: World Scientific)

[3] Bahr C and Kitzerow H-S 2001 *Chirality in Liquid Crystals* (Berlin: Springer)

[4] Collings P J and Hird M 2017 *Introduction to Liquid Crystals: Chemistry and Physics* (Boca Raton, FL: CRC Press)

[5] Davis A L, Brink D J, Scholtz C H, Prinsloo L C and Deschodt C M 2008 Functional implications of temperature-correlated colour polymorphism in an iridescent, scarabaeine dung beetle *Ecol. Entomol.* **33** 771–9

[6] Demus D, Goodby J W, Gray G W, Spiess H W and Vill V 2011 *Handbook of Liquid Crystals, Volume 2A: Low Molecular Weight Liquid Crystals I: Calamitic Liquid Crystals* (John Wiley & Sons)

[7] Fábián B, Sega M, Voloshin V P, Medvedev N N and Jedlovszky P 2017 Lateral pressure profile and free volume properties in phospholipid membranes containing anesthetics *J. Phys. Chem.* B **121** 2814–24

[8] Gabriel J-C P, Sanchez C and Davidson P 1996 Observation of nematic liquid-crystal textures in aqueous gels of smectite clays *J. Phys. Chem.* **100** 11139–43

[9] Lazaris A, Arcidiacono S, Huang Y, Zhou J-F, Duguay F, Chretien N, Welsh E A, Soares J W and Karatzas C N 2002 Spider silk fibers spun from soluble recombinant silk produced in mammalian cells *Science* **295** 472–6

[10] McKenzie I, Cammidge A N, Dilger H, Gopee H, Scheuermann R, Stoykov A and Jayasooriya U A 2010 Muon spin spectroscopy of the discotic liquid crystal HAT6 *Phys. Chem. Chem. Phys.* **12** 9900–8

[11] Michot L J, Bihannic I, Maddi S, Funari S S, Baravian C, Levitz P and Davidson P 2006 Liquid–crystalline aqueous clay suspensions *Proc. Natl. Acad. Sci.* **103** 16101–4

[12] Neville A and Caveney S 1969 Scarabaeid beetle exocuticle as an optical analogue of cholesteric liquid crystals *Biol. Rev.* **44** 531–62

[13] Tahir H M, Zahra K, Zaheer A and Samiullah K 2017 Spider silk: an excellent biomaterial for medical science and industry Punjab Univ *J. Zool.* **32** 143–54

[14] Trudell J, Hubbell W and Cohen E 1973 The effect of two inhalation anesthetics of the order of spin-labeled phospholipid vesicles *Biochim. Biophys. Acta Biomembr.* **291** 321–7

doi:10.1088/978-0-7503-1362-9ch10

www.ingramcontent.com/pod-product-compliance
Ingram Content Group UK Ltd.
Pitfield, Milton Keynes, MK11 3LW, UK
UKHW051848100926
8556IPUK00005B/523

9 780750 313636